D1717654

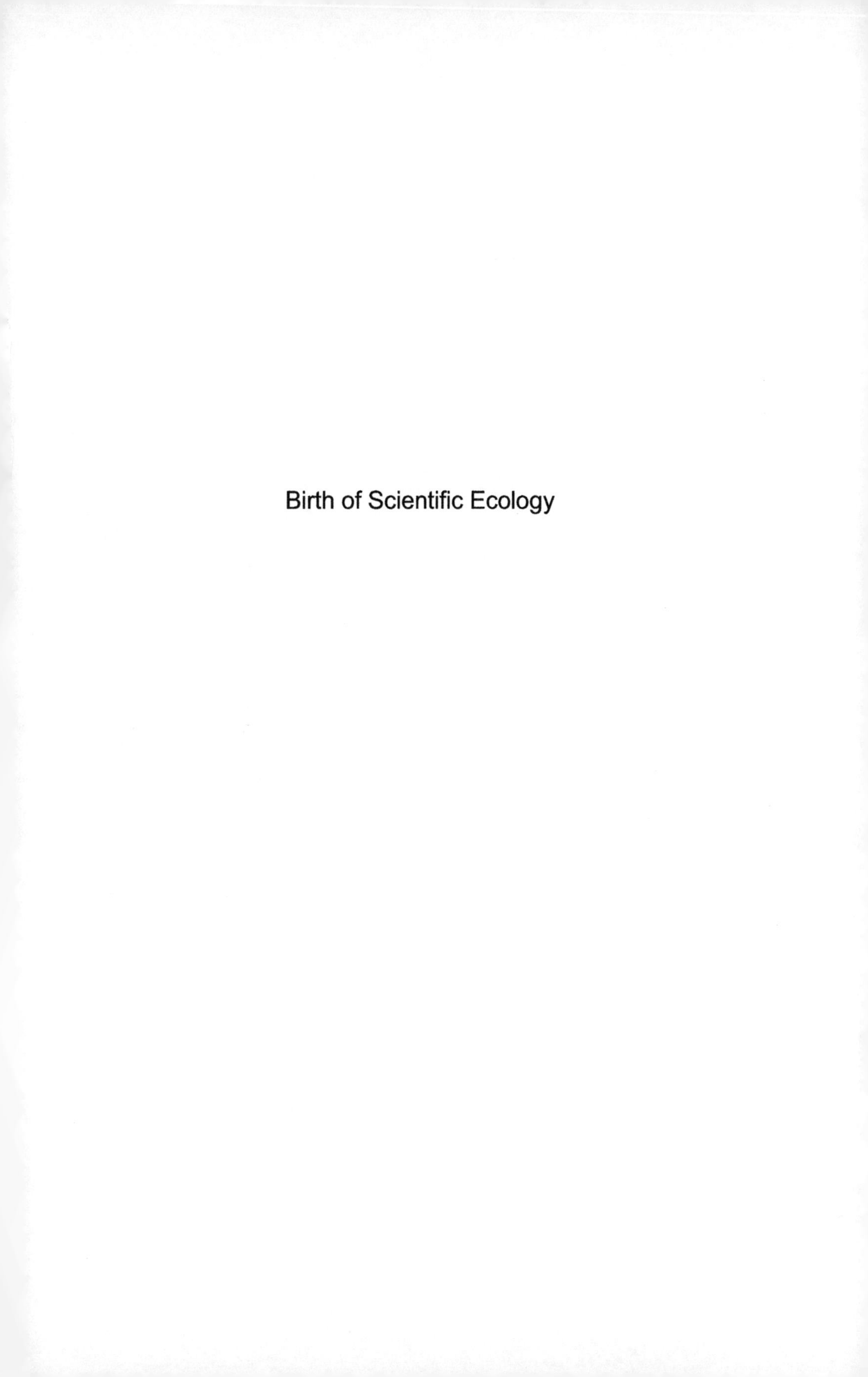

Birth of Scientific Ecology

Series Editor
Françoise Gaill

Birth of Scientific Ecology

Eugenius Warming
(1841–1924)

Patrick Matagne

WILEY

First published 2024 in Great Britain and the United States by ISTE Ltd and John Wiley & Sons, Inc.

ISTE Ltd
27-37 St George's Road
London SW19 4EU
UK

www.iste.co.uk

John Wiley & Sons, Inc.
111 River Street
Hoboken, NJ 07030
USA

www.wiley.com

Library of Congress Control Number: 2023947134

British Library Cataloguing-in-Publication Data
A CIP record for this book is available from the British Library
ISBN 978-1-78630-929-7

Contents

Foreword

In France, studies specifically devoted to the history of scientific ecology are recent, and their first authors can be counted on the fingers of one hand. The small group they formed was made up, in alphabetical order, of Pascal Acot, Jean-Paul Deléage, the late Jean-Marc Drouin (1948–2020) and Patrick Matagne, author, among others, of this book. All of them have published several acclaimed works, and have developed mutually fruitful and often esteemed scientific relationships.

Patrick Matagne is an outstanding historian of science. We only have to read his books and publications to be convinced. He holds a master's degree in history commenced under the supervision of Alain Corbin, then a doctorate in epistemology and history of science (on the history of scientific ecology) after defending a thesis on this theme in 1994 under the supervision of François Dagognet (1924–2015). He then became a lecturer at the IUFM Nord-Pas-de-Calais, then Poitou-Charentes Poitiers.

This researcher with vast knowledge is also a man of the field who knows what he is talking about, as he also holds a master's degree in Natural Sciences from the University of Poitiers. In 1998, after working on the Mesoamerican biological corridor, these degrees and scientific qualities led him to lead a seminar on the history of ecology at the University of Costa Rica's Faculty of Social Sciences.

His latest work cannot be reduced to a simple biography, however erudite, of the Danish biologist Eugenius Bülow Warming (1841–1924), the botanist most often regarded as the founder of *scientific* ecology. This book is far more important, because when we look at Patrick Matagne's work, we discover that its author combines all the scientific and stylistic qualities of his previous works.

Thus, the book resulting from his thesis entitled *Aux origines de l'écologie* and subtitled "*Les naturalistes en France de 1800 à 1914*" (CTHS, histoire des sciences et des techniques, 1999) presented remarkable originality: that of considering the importance of provincial naturalists, often non-professionals, in the emergence of modern ecological problems and, conversely, in the way they integrated into their "local" reflections the work of the first great ecologists. As a result of this book, the origins of scientific ecology, hitherto dated to the second half of the 19th century, can be traced back to the 1800s. This result is far from insignificant, which is no mean feat for a first major publication.

Similarly, the brilliant and profound introduction to *Comprendre l'écologie et son histoire* (Delachaux and Niestlé 2002) focused on the first North American nuclear explosion in the New Mexico desert in 1945. To my knowledge, Patrick Matagne was one of the first to highlight its tragically paradoxical importance, which ushered humanity into the "ecological age" by emphasizing the fragility and *vital* importance, in the truest sense of the word, of ecosystems.

This is why *Les Enjeux du développement durable*, the collective work he edited and published by L'Harmattan in 2005, does not represent a kind of editorial parenthesis in the reasoned publication of his research, but rather a fruitful transition. Indeed, the proceedings of the study days organized in 2003–2004 by the Espace Mendès France (Poitiers) mark a decisive stage in his work. Edgar Morin, who wrote the foreword to the work, made no mistake: this collection has contributed to bringing to scientific ecology what politics is most challenging for the future of the planet's civilizations.

It is therefore hardly surprising that *La Naissance de l'écologie* (Ellipses 2009), which focuses on the work of Eugenius Warming as a criterion for the scientific nature of this new branch of biology, is now inseparably perceived as describing not only the birth of a new science, but also of a new political awareness, if not soon a new morality. This latest book from Patrick Matagne has inherited the same qualities as his previous works.

The book's remarkable introduction takes the reader by the hand (the author is no stranger to this): "Eugenius arrived at his destination at around ten o'clock on July 8, 1863. After spending the night at Manoel's farmhouse, his guide since his arrival in Rio de Janeiro, they left the mules to travel the last few leagues still separating them from Lagoa Santa on horseback." It sounds like the beginning of an adventure book, even if it is the beginning of a highly documented, wonderfully illustrated academic history of scientific ecology. And, as in an adventure book, the reader, trapped by curiosity, wants to know more and more about Eugenius. In this respect, the use of first names introduces a welcome complicity between the biographer and his readers. Until now, we thought we were simply discovering a major founder of ecology, and that is what happens as we read on. But this founder is a human being, so essential that we often wish we had known him.

In this way, Patrick Matagne gives us a useful lesson in epistemology: despite what a superficial glance may suggest, science is never neutral, but always developed by sensitive human beings, in given material and cultural conditions. When he describes Warming's distress at learning of his mother's death in Lagoa Santa three months after leaving her, he is in no way concerned with anecdote, but paves the way to an innovative method of practicing the history of science, without which we would miss important aspects of the great Danish ecologist's thought.

Pascal ACOT
Doctor
Institut d'histoire et de philosophie des sciences et des techniques
CNRS, Université de Paris 1, ENS-ULM

Acknowledgments

Pascal Acot, a French pioneer in the history of scientific ecology and a historian of environmental sciences, climate and climatology, did me the honor of reviewing my text, with all the wisdom and high level of expertise that characterize him.

Jean-François Beauvais, a temperate and tropical botanist, whose analytical eye supported the writing of this book, was kind enough to mobilize his naturalist skills.

My gratitude to them goes beyond mere convention. Their encouragement has played no small part in the success of this work, for which I am – of course – solely responsible.

Introduction

Eugenius arrived at his destination around ten o'clock on July 8, 1863.

After spending the night at Manoel's farmhouse, his guide since his arrival in Rio de Janeiro, they left the mules to travel the last few leagues still separating them from Lagoa Santa on horseback. Eugenius had planned to stay there for two years, and would stay for an additional year.

"The morning was pleasant. Bluebell-shaped flowers and many others adorned the hills"; "the dew was like pearls in the grass", he wrote in his diary[1]. At almost 800 m above sea level, on the vast plateau of what is now Minas Gerais, the temperature was mild at that time of the year[2]. "I let my gaze wander over the large square in the center of town".

Although the square was vast, Eugenius was soon to discover that the "town" was no more than a "miserable village" with low-slung houses and streets made of earth, limestone and short grass, as his photographs show. In the early 19th century, Lagoa Santa had 500 inhabitants and 80 houses. Manoel pointed out the one he needed to go to. He entered and stood waiting while his guide went to inform the owner of the premises who was resting in his garden at the back. After his morning stroll, he usually enjoyed the shade of the biribá and palm trees[3].

---

1 Pages 63–73 from Eugenius Warming's diary are published by Klein, A.L. (2000). *Eugen Warming e o cerrado brasileiro*. UNESP, São Paulo. Other extracts are taken from Prytz, S. (1984). *Warming Botaniker og Rejsende*. Bogan, Lynge.

2 22°C annual average. The coldest month is July (average 18.8°C); the warmest is February (average 23.7°C).

3 *Rollinia laurifolia* (Annonaceae), *Acrocomia sclerocarpa* and *Cocos capitata* (Arecaceae).

Figure I.1. *Dr. Lund's house and garden, on the right in the image (public domain via Wikimedia Commons)*

Eugenius was nervous. He was about to meet the Danish scientist whose secretary he was to become. Then entered "a thin man with gray hair". Peter Wilhelm Lund had passed the age of 60, Eugenius was not yet 22.

"To my surprise, he greeted me in German[4]. He was normally supposed to use this language with Mr Brent, who replaced me temporarily. I think I replied in German, but then he realized he would have to speak in Danish". We can assume that Eugenius' surprise must have been tinged with a certain amount of annoyance, given his strong patriotic feelings against the conquering German Confederation.

For the time being, dramatic news awaited him.

"After a few minutes of conversation, he [Lund] remembered that mail had arrived for me"; "the first letter I opened, with a strange sense of anguish, brought me paralyzing news: my mother was dead". We can imagine the grief of the young man, an only child and fatherless, discovering that his mother had died on May 5, 1863, less than three months after his departure.

He landed in Rio de Janeiro on April 27 and stayed for five weeks. Far from Europe and his University of Copenhagen, he was delighted to discover the tropical nature surrounding the city. Like all naturalists, he observed, collected, drew, described and – a rarity at the time, given the technical difficulties that discouraged many beginners – used the bulky and fragile camera he had packed in his luggage

4 "*Ah, es ist Herr Warming, setzen Sie sich gefälligst Nieder*" (Ah, it's Mr Warming, please sit down).

(Davanne 1867; Gunthert 1999, p. 205). Cautious and organized, he took a two-week photography course before his departure.

His uneasiness, and even his guilt, can be seen in the lines left in his diary:

> It's true that I had received a letter in Rio from my mother's brother informing that she was ill, but, as he himself had said, there was no danger, as I had left her in very good health and as for many years she had never been ill, I didn't attach much importance to the fact.

His uncle, who took him and his mother in after his father's death when he was barely three, probably did not want to alarm him. In any case, if Eugenius had decided to return, he would have had to face another arduous journey by mule to Rio de Janeiro, wait for a ship to take him to Europe and sail for many weeks.

A strange coincidence: Lund also learned of his mother's death while on a study trip in Italy with the Danish botanist Joakim Frederik Schouw. Arriving in Sicily, they hired a cart and two mules, passing through Messina, Catania, Syracuse and Agrigento. In Palermo, Lund received a bundle of letters, including one from his cousin telling him the sad news. He returned to Copenhagen for the last time in the summer of 1831. With no family ties, he hesitated between settling in Paris or Brazil. According to Danish zoologist Johannes Theodor Reinhardt with whom he had an ongoing correspondence, his mind was already made up (Luna Filho 2007, p. 71 *ff*).

What does the face of young Eugenius, photographed at the age of 21 shortly before his departure for Brazil, express?

The half-length portrait shows a serious, almost austere face, with a bare forehead and hair swept back. He is wearing glasses, a pencil beard and a budding moustache. Photographs from this period have a certain frozen quality, due to the technical necessity of requiring the subject to remain motionless. This portrait, taken in a photographer's studio, has been retouched like almost all of them. The background is neutral, with only the upper part of the torso visible. He did not stare at the lens, his gaze seemingly lost in contemplation of a distant horizon.

The intention here was not to show the subject in a particular situation, unlike those naturalists captured in postures that give the illusion of movement while in their study, sometimes in nature observing a detail, magnifying glass in hand. Later, after a brilliant career, Professor Warming came to take his place in these "galleries of contemporaries", "photographic portraits of famous figures from politics, science and the arts", fashionable from the second half of the 19th century onwards (Rouillé and Marbot 1986, p. 33; Gunthert 1999, pp. 13–14)[5]. These representations were intended to signify the social success or scientific notoriety of the person being "portrayed". The physiognomy of the Swiss Johann Caspar Lavater was popular. Portraits were thought to reveal the personality, feelings emotions and even the soul of the subject.

Figure I.2. *Portrait of Warming at age 21 (Klein 2002, p. 19)*

When did Johannes Eugenius Bülow Warming conceive the project that would make him the author of the first treatise on ecology? What was it intended for?

Mr Launay, professor of history at the Université François Rabelais in Tours, warned his students: to shed light on an individual's intellectual, ideological, political and spiritual journey, ask yourself where he was, what he was going through and what decisions he made when he was 20.

This is how the destiny of an individual would then be written.

No one better than Balzac knew how to play with the destiny of his characters. In *La Comédie humaine*, he takes up the biological notion of milieu, defined by Comte in his *Cours de philosophie positive* and extended to human societies, where individuals interact with one another. Balzac sought to grasp the laws governing the distribution of social species. He analyzed the decisions, behaviors and aspirations of his characters. Drawing on the animal nomenclature established by paleontologist

5 At the end of the 19th century, a snapshot was gradually mastered thanks to the introduction of the gelatin–alkaline combination.

Cuvier and zoologist Buffon, he aligned the physical and the moral (Cohen 2004; Matagne 2004; Collet 2019).

Eugenius was born on November 3, 1841, on the small Danish island of Mandø in the Wadden Sea, the only son of Lutheran minister Jens Warming and Anna Marie von Bülow. Following his father's untimely death, his mother left the island to live with her young child on the east coast of the Jutland peninsula, near Vejle, where Eugenius attended school before completing his secondary education in Ribe, less than 20 km from his native island. Introduced to botany by a natural history teacher, he became familiar with the plants of the Jutland coastline.

He enrolled at the University of Copenhagen in 1859, the publication year of Darwin's *On the Origin of Species*. An opportunity to travel presented itself. Professor Reinhardt proposed that he leave for Brazil to become secretary to the zoologist and paleontologist Lund, whose assistant he had been. Eugenius interrupted his studies.

He left on February 17, 1863 and returned to Denmark in October 1866.

On his return, he completed his studies in Copenhagen, then moved to Munich and Bonn to continue his research. He defended his doctoral thesis in 1871, the same year he married – he had eight children – then became temporary assistant professor at the University of Copenhagen at the age of 32, and professor at the University of Stockholm from 1882 to 1885. He returned to his home university, where he taught until his retirement. He was also the director of the Botanical Garden, where his herbarium, drawings, photographs and diary are now kept. A pedagogue, he devoted himself zealously to teaching, publishing botanical textbooks that met with great success. A man of the field, he felt it necessary to take his students outside the walls of the university. To open them up to the concepts and methods of botany and plant ecology, the botanical garden was not enough.

His work was enriched by his travels, which brought him into contact with landscapes and flora from latitudes as diverse as Greenland, Venezuela, the Caribbean, the Faroe Islands, Scandinavia and Tunisia, not to mention short stays in the Alps and the south of France.

How did the opportunities, the ups and downs of a life spanning more than 80 years shape the pastor's son, who went on to become an internationally renowned scientist? What role did the unpredictable, the unexpected and the uncertain play?

To shed light on the work and the man, should we highlight his father's death before he could retain any conscious memory of it, but which led to his move to the continent? Or his encounter with a natural history professor during his secondary school years? His decision to leave university before completing his studies and accept the offer to cross the Atlantic to become the secretary of an old scholar? The news of his mother's death, far from everyone and everything? The discovery of exotic flora whose physiognomy, so different from that of Denmark, turned his understanding of botany and botanical geography upside down; the great European conflicts that severely affected his homeland?

These questions, along with others, will run through this narrative, whose ambition is to make the life of Johannes Eugenius Bülow Warming intelligible a posteriori, although it was unpredictable a priori, like all human lives (Morin 2021, p. 29–46, p. 143).

PART 1

From Mandø to Lagoa Santa

1

Eugenius' Birthplace

1.1. A deprived small island

A traveler discovering the island of Mandø in the 1840s on a November day with low clouds heralding rain would not have had the most pleasant of impressions. Certainly, temperatures would be above zero all day long, but they would remain in the single digits.

Coming from the nearby mainland at low tide through a passage created by the sand bank and roughly protected by two rows of posts – which have to be rebuilt every spring – visitors would discover a hostile, wind-beaten environment, a low, marshy land. According to an 1839 map, the altitude did not exceed 5 m above sea level. The day-tripper was not to linger too long, as the low light would fade by 4 p.m. An unpaved road took you to Ny Mandø (New Mandø) in just a few minutes, where the first houses in the mist could be made out and, behind them, the church built outside the village, as tradition dictates. Visitors could then walk between the houses protected by the dunes, with their adjoining garden surrounded by hedges on the south side.

It was here that Eugenius was born on November 3, 1841, in the land of "eternal November", in the words of Danish poet Henrik Nordbrandt. For him, in Denmark:

> The year has 16 months.
>
> November, December, January, February,
>
> March, April, May, June, July,
>
> August, September, October,

November, November, November, November (Nordbrandt 1986; Viegnes et al. 2020, p. 151).

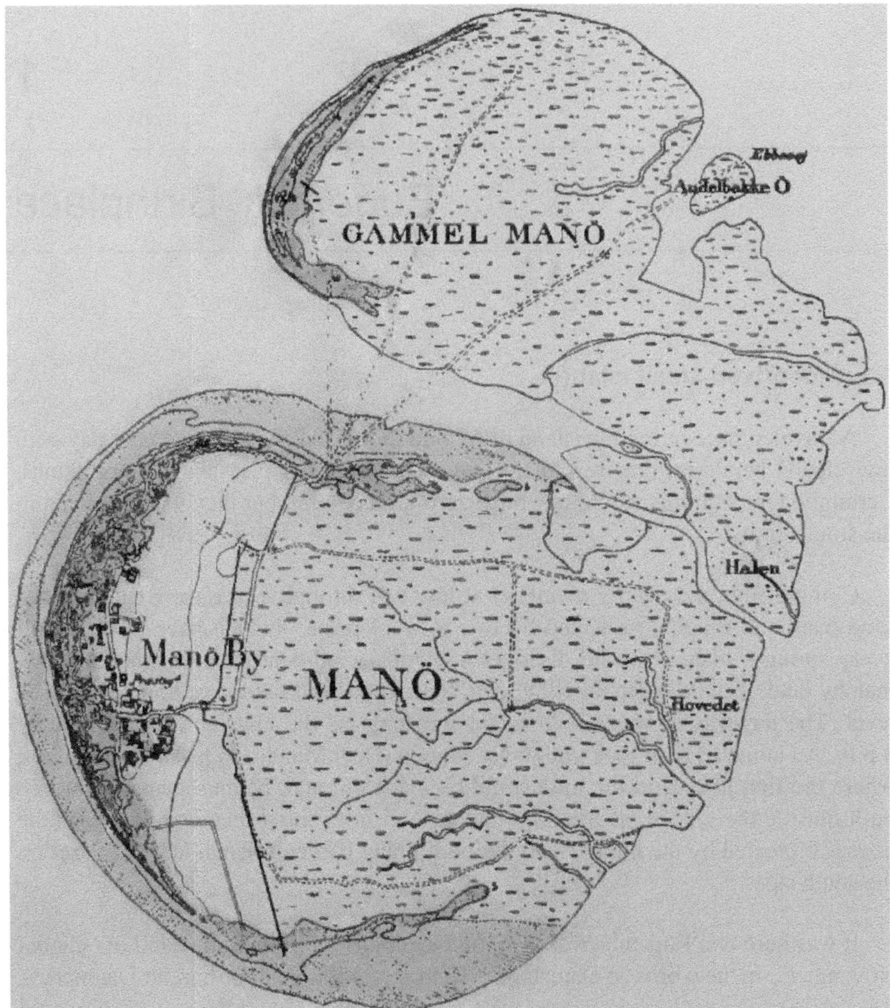

Figure 1.1. *The island of Mandø (Manø) around 1870 (copy of the bulletin of general staff) (Warming 1906–1919, p. 116)*

Mandø, Denmark's smallest island, is part of the northern archipelago of the Friesland Islands, which form a dividing line between the North Sea and the Wadden Sea. The island is separated into two by a channel. By the time of

Eugenius' death in 1924, the two parts had virtually fused together due to the relatively rapid silting up and filling in of the channel (Oorschot 2009)[1].

As an object of study, Mandø has long been deserted by historians and archaeologists for whom "the general idea is that nobody could ever have been so stupid as to live there, except if you were desperate, a monk or on the run" (Oorschot 2009, p. 3). The primary function of this desolate island would have been to house the graves of shipwreck victims.

The name Mandø first appeared in the literature in 1231. A church was mentioned in 1325. It was stated that it must pay four skilling sterling (silver coins) to the diocese of Ribe, on which it depended. There were three churches in the early 15th century. The present building dates from 1639, built on the highest point for safety. It was restored in 1727. Meanwhile, Lutheranism had implanted itself in the kingdom of Denmark–Norway under the reign of Frederick I. His son, Christian III, made the Reformation the national religion, and the University of Copenhagen became Lutheran. Royal property, the island of Mandø was bought at auction by its inhabitants in 1741. Traditionally, women were farmers on the island, in difficult conditions, as coastal marshland extends into arable land. Men were fishermen. Sailors, helmsmen, captains, they left their wives alone with their children, sometimes for weeks at a time.

The first school was built in 1776, with a home for the sole teacher. In 1884–1885, it was enlarged with the opening of a second class; by the end of the century, the island had a population of around 250 (92 farmers, 59 sailors, 22 industrialists and 13 tradespeople). There was a merchant, an innkeeper, two blacksmiths, a baker and a carpenter. An 1890 census counted 262 inhabitants but only 37 in 2013.

From 1870 onwards, the inhabitants tried to fix their island's dune complex with vegetation to protect them from recurrent flooding. To the east, a dike was built by the women in 1887 to protect their village. The same year, an unpaved road linked the island to the mainland, passable at low tide and outside of storms. By 1888, the dike ran all the way around the island, except to the south-west, where the village was protected by the dunes.

Women were the true administrators and custodians of their territory, against all odds. We imagine them as strong and pugnacious, guarantors of the durability of a people and a village built, destroyed, moved, rebuilt on unstable, shifting ground,

1 Maps by Mejer (1651), Sørensens (1696, 1794, 1839, 1848, 1861, 1870, 1901 and 1910).

scanning the horizon beyond which the men disappeared and did not always return. The graves of the shipwrecked bear witness to this (Valdemar 1231; Geffroy 1851; Domeier and Haack 1963; Carré 1976).

The literature of the early 20th century liked to link the land to its inhabitants, in the tradition of German anthropogeography. The character traits of the Jutlanders, descendants of the ancient Cimbres people, are said to have been forged by the harshness of the farmers' toil and the fishermen's life, a harshness determined by that of the climate. The historical, geographical and economic studies on Denmark, published in particular to coincide with the 1900 World Fair, take up what has become commonplace, namely the opposition between the "alert" Fionian (from Hans Christian Andersen's native island) and the "cheerful" Selandese (on the larger island of Copenhagen), and the "overbearing and serious" Jutlanders, whether continental or insular (Denmark has over 400 islands) (Paul-Dubois 1909, p. 659).

What can we learn from Gammel Mandø (Old Mandø) and its tragic history? asks Leo Oorschot in a historical and archaeological study of *The Flooded Village of Mandø* (Oorschot 2009, p. 3).

In the 20th century, studies were being carried out into the consequences of the appalling storm of October 11, 1634, which washed away the dikes and ancient protection built in the 8th and 9th centuries. The "great flood" accompanying this storm led to the destruction of many villages on Denmark's west coast, between Ribe – opposite Mandø – and Tønder, 50 km to the south. On this terrible night, 500 people are thought to have perished. Ribe was reported to have a flood 6 m high. Gammel Mandø was reportedly unable to stand firm (Pontoppidan 1763–1781; Domeier and Haack 1963).

It is possible that the inhabitants took refuge in the dunes to the southwest of the island and set about building New Mandø (Ny Mandø), where Eugenius was born. But catastrophic floods were common at the time. It is not certain whether the flood of 1634 was the final blow to the village. Several authors mention another large flood in 1558, which caused half the population to flee. Those who remained escaped death by taking refuge on the roofs of their farms. After the tragedy, Svend, the island's first pastor, came to support the few families still clinging to their small island and decided to build Ny Mandø in the southern part of the island, protected by dunes and relatively higher above sea level. The story goes that the pastor did not hesitate to come from the mainland even when the ground was flooded. It became customary to say of someone who could make progress where others were sinking in the mud: "he must have Mr. Svend's boots".

1.2. A Lutheran pastor's son

When Eugenius' parents moved to Ny Mandø, they were neither desperate nor on the run. His father, Jens, was born on March 25, 1797 in the hamlet of Råhede (parish of Hviding), on the west coast of Jutland, less than 10 km from the island where he became pastor. His parish came under the authority of the diocese of Ribe.

In Denmark, even more than in other Scandinavian countries, the parish is the true home of the religious community, in a country where the Lutheran Church plays a leading role (Jolivet 1961, pp. 24–26). It built a Christian ethic based on sacred texts, imbued with rationalism, and advocating altruism and charity. Originally, it was traditionalist and conservative in political and social terms. One thinks of the austere, authoritarian figure of the patriarch who reigns over his community, controlling its spirituality and morals, in the Danish film *Babette's Feast*, directed by Gabriel Axel (1987), based on a short story by Karen Blixen. The action takes place in a small village in Jutland in the 19th century (Blixen 1961). This pastor would be quite comparable to the father of the main character in Danish writer Henrik Pontoppidan's *A Fortunate Man* (*Lykke-Per*), written between 1898 and 1904.

At the time of Eugenius' father's ministry on the island of Mandø, "Grundtvigianism", which relied more on the living word handed down by the apostles than on sacred texts, claimed to be the work of Pastor Grundtvig. A Lutheran conservator turned reformer, fighting against the religious rationalism of the Enlightenment, Grundtvig believed that religious life should animate and direct our actions (Lehmann 1931). The pastor who became *A Fortunate Man's* stepfather is portrayed by Pontoppidan as more humane and attentive to individual feelings and aspirations than his character's father[2].

Grundtvig preached for the restoration of a religious and national sentiment, "after a phase of contraction and withdrawal" (Mougel 2006, p. 6) of Denmark, due to the loss of Norway to Sweden under the Treaty of Kiel signed on January 14, 1814. Involved in politics, he obtained a ruling from the Upper House that communities should choose their pastor. Legislative power was then shared between the Crown and the *Rigsdag*, Denmark's two-chamber parliament (1849–1953). Members of the conservative Upper House (*Landsting*) were appointed by a complex censal system (12 out of 66 members were appointed for life by the Crown). Grundtvig sat in the reformist Lower House (*Folketing*), whose members

2 Another militant current, *Indre mission* (interior mission), urges believers to evangelize, like the missionaries in the colonies.

were elected by universal male suffrage (must be over 30). He initiated a major popular education movement.

In bankruptcy, the Danish colonial empire sold almost all its trading posts and colonies in Africa and Asia to Great Britain. The most traumatic event was the loss of the Duchies of Lauenburg, and above all Schleswig and Holstein, at the end of the two Duchy Wars. The first (the Three Years' War, 1848–1851) pitted Denmark against the German Confederation, the second against the Kingdom of Prussia united with the Austrian Empire, from February to July 1864. Denmark, which had believed it could benefit from the support of Sweden and Norway, found itself alone in the face of an invasion without a declaration of war. It was a setback for the Copenhagen government, opposed to the "Eider Danes", who wanted only to keep Schleswig danophone, the river Eider marking the new German–Danish border, considering the claims of the German nationalists legitimate.

Meanwhile, the Constitution of June 5, 1849, signed by King Frederick VII, made Lutheran Protestantism the national church, but gave religious freedom to the Danish people. However, the King and Queen had to belong to the Evangelical Lutheran Church (article 6). The Danish "People's Church" was supported by the state, although this did not exclude support for other faiths or denominations (Dübeck 1997).

Under the terms of the Treaty of Vienna of October 30, 1864, which put an end to the Second Duchy War, the Danish border was now partly marked by the Kongeå, the river that historically formed the administrative boundary between the northern and southern regions. Schleswig was administered by Prussia, Holstein by Austria. Following Prussia's victory over Austria at the Battle of Sadova (July 3, 1866), the two duchies were integrated into Prussia as the province of Schleswig–Holstein, part of the North German Confederation.

German blood ran in Eugenius' veins. His mother, Anna Marie von Bülow af (de) Plüskow-Aggerupgaard-Bjørnemosen, was born in Kolding on November 23, 1801, 25 km south of Vejle. An analysis of her full name, in the order in which it is written, reveals her genealogy. She descended from an old noble family known from at least the 13th century. Bülow is located in Western Pomerania, in a region that became the Grand Duchy of Mecklenburg-Schwerin between 1815 and 1918. The Bülow and Plüskow lines joined forces. Aggerupgaard (Agerupgård) is a large manor farm, with associated farms. There are several hundreds of these throughout Denmark. They were owned by aristocrats who enjoyed tax privileges until the early 20th century. Agerup is located in the parish of Våbensted, in the municipality of Guldborgsund (on the island of Lolland, Denmark's fourth largest). In the early

18th century, the property was leased to Lieutenant-Colonel Nicolaj Christoph von Bülow af Plüskow. Christian Friederich von Bülow was born here on March 24, 1724. Finally, Bjørnemosen (Odense) refers to another farm acquired by Adam Gottlob Josva von Bülow, then by his son Rasmus Hartvig von Bülow, in the first half of the 19th century.

Eugenius' maternal branch gave Denmark illustrious characters, close to power, some of whom acquired Danish nationality and were admitted to the nobility of their host country, where they made their fortunes. In fact, following his mother's death, Eugenius received an inheritance that enabled him to pursue his studies in peace, before settling comfortably with his family and household staff in one of Copenhagen's best neighborhoods[3].

Eugenius' future father, Jens Warming, was a farmer's son, the second of nine siblings. His parents owned a farm in Ribeegnen. He was the only one in the family to have studied. He must have attended the Lutheran church in his parish of Hviding. In January 1833, he was chaplain at Lejrskov-Jordrup, two Lutheran churches located 10 km and 15 km west of Kolding, then at the end of September 1837 at Nørup-Randbøl, less than 20 km west of Vejle, where he built himself a small house with a plot of land. Jens practiced in the area where Anna Marie lived. He married her on October 12, 1838. He became pastor in Mandø on January 31, 1841, where Eugenius was born on November 3.

After Jens died of pneumonia on December 15, 1844, his widow lived for a time in Nørup with her brother on his farm (Nørupgård), some 10 km from Kolding. From 1854, she lived permanently in Kolding with her mother, Johanne Jensdatter Wissing, wife of von Bülow. Widowed in 1835, she came from a family of wealthy Kolding merchants and had a large fortune (Prytz 1984).

Eugenius rarely returned to his native island, no doubt deterred by political instability and wars. Lund put forward this argument in his decision not to return home. Between the end of the Duchy Wars and 1918, access to the island of Mandø was not easy. In May–June 1916, the largest naval battle of World War I, fought in

3 Johan Bülow made a career in the army, became owner of a large farm (Sanderumgård), chamberlain and then field marshal to the future King Frederik VI; Bernhard Ernst von Bülow held both Danish and German nationalities, and became Foreign Minister; Adolf von Bülow married the daughter of a Danish soldier and joined the Ministry of Finance; Lieutenant General Frederik Rubeck Henrik von Bülow led the Danish army against the Prussian army at the Battle of Kolding on April 23, 1849, an episode in the First Duchy War. The Danes retreated to Vejle and Fredericia, where their leader has had a statue since 1869. See: https://www.ancestry.com; https://www.geni.com, https://www.myheritage.com; https://finnholbek.dk; https://www.geneagraphie.com; https://www.olhus.dk; https://biografiskleksikon.lex.dk/.

the Jutland sea area, pitted the Royal Navy against the German navy. The coastal area was mined, from the mouth of the German river Ems in the south to the Danish port of Esbjerg in the north. However, from the 1880s onwards, Professor Warming made a number of excursions to the coast and islands of West Jutland and rediscovered his native island, which he visited for the first time in 1889. After 1895, he stopped making long trips, but his students praised his robustness and perseverance in going out into the field with them. Part of his work was devoted to flora and vegetation, which enabled him, as a teenager, to make his first observations of the ways in which plants adapt to a particularly selective environment.

2

Jutland: At the Roots of a Passion

2.1. Decisive choices

Since 1814 (law on schools for the people, *almueskolelov*), education has been obligatory for all children aged 7–14 years old, with parents choosing between private tuition, private or state schools (Appell 1948).

Eugenius attended public school in the village of Nørup, 20 km west of Vejle. He lived with his mother on her brother's farm, Jens Vissing Conrad von Bülow[1]. In this rural area, Eugenius began to nurture a passion for nature, then represented by the Randbøl moor. Before he went to school, his mother sent him to drawing classes and let him learn German, geography and natural history. Between 1850 and 1856, he continued his education at the Latin school (*latinskole*) in Kolding, a port town some 30 km south of Nørup, whose streets sloped down to the fjord (Appell 1948; Rauser 2004)[2]. Eugenius said he was confronted by the geographical proximity of the conflict – the First Duchy War was drawing to a close – and the patriotism of the Danes in Kolding. He remembered assiduously frequenting the school library and rubbing shoulders with country folk from a background different from his own.

For high school, his family chose the town of Ribe, some 50 km west of Kolding. Ribe, the oldest town in the Kingdom of Denmark, was a trading center founded in Viking times, largely destroyed by fire in 1580 and rebuilt around 1600. By the time Eugenius took up residence here, it had lost its economic importance.

---

1 Eugenius' uncle married Frederikke Lovise Eiler on November 12, 1856. Their first child was born in September 1857, the second in November 1859. Two more children followed in 1861 and 1862.

2 Other schools: city schools (*borgerskoler)*, country schools (*friskoler*) and *højskoler* for students over 18 years old.

From then on, “the city vegetated obscurely in the shadow of its cathedral”, wrote historian Lucien Musset (1948, p. 316).

Figure 2.1. *View of the Ribe cathedral tower looking west over the meadows. Photo by Warming (1906–1919, p. 108)*

The private high school where Eugenius was enrolled had a very good reputation and aimed for excellence. It prepared students for entrance to the University of Copenhagen, the only university in the country until the Aarhus University was founded in 1928.

Eugenius’ family had ties and memories in this diocese. As we have seen, Ribe was a diocese on which Eugenius’ father’s parish depended. In addition, part of the family of Frederikke Lovise Eiler, his maternal uncle’s wife, lived in the diocese of Ribe. Her father was a minister in the parish of Jernved (he died in 1845), about 15 km from Ribe. However, Eugenius’ arrival at the age of 15 was the direct consequence of a political and educational choice made by Denmark, which his family had to face up to. An ordinance of April 17, 1839 led, in the years that followed, to the gradual abolition of small Latin schools, concentrating resources and pupils in the larger ones. Finally, his mother may have avoided choosing Copenhagen because of the recent memory of the terrible cholera epidemic of 1853, which claimed 4,800 lives. In the years 1856–1857, sanitation work and the reorganization of certain activities and neighborhoods were underway: a new livestock market, a safer municipal water supply, construction of healthy (social) housing outside the city center (Brumleby, Østerbro).

At Ribe high school, encounters and a context, notably linked to the consequences of the First Duchy War, were to be decisive at an age when personality was being built. Eugenius also developed a passion for botany and the landscapes of West Jutland.

2.2. Bewitching moorland

Eugenius usually walked along the coast of Jutland (*Jylland*), which he could reach on foot from the town. Its channel had gradually silted up since the terrible flood of 1634. It affected the entire coastline, but spared Esbjerg to the north, making it Denmark's main direct outlet to the west.

The high school student recorded a particularly memorable excursion in his diary:

> One afternoon, my comrades and I were taking a walk on the beach. There we were, by the water, as the luminous disk descended from the cloudless sky to the sea – then thousands of seabird voices formed a strangely mystical chorus. Twilight fell over the waters, amplifying a sense of mystery and secrecy. In a very special atmosphere, which I'll never forget, we returned to Ribe via the polder meadows, which made walking difficult, as we had to cross large ditches.

In his diary, he repeatedly recalls the strong impressions left by this episode, arousing a mystical exaltation and a desire to penetrate the secrets of nature.

"Central and western Jutland was, at the beginning of the 19th century, nothing but a sea of moors, as far as the eye could see" (Musset 1948, p. 316). The coast and the western moorland could seem hostile to the traveler coming from latitudes with a less harsh climate, a less unstable coast, and more varied vegetation and landscapes. Despite the oceanic influence, the shores could be frozen for long weeks in winter. Born in "the poorest of the parishes of Jutland, in the middle of the moor" (in Sædding), further north, the great writer Kierkegaard only twice visited this moor of sand and heather, hot in summer and freezing in winter. An oil on canvas painted in 1855 by Frederik Vermehren, *Shepherd on the Heath* (*Fårehyrde på heden*), shows a shepherd with his dog in a flat, monotonous landscape where sheep graze. In the foreground is heather.

Eugenius did not give a description that would dissuade the walker during this escapade of high school students experiencing freedom. Like the Jutland writer and pastor Steen Steensen Blicher, who grew up in the marshy areas of Central Jutland, he was under the spell of this land, the historic heart of Denmark. Blicher describes this land of legends, the land of his ancestors (the Jutes people, known in the 5th century) and this landscape, for him also a symbol of peace and freedom:

> This moor you're entering now is vast and flat as the sea, and here and there, on the horizon, you see churches and houses, as if jutting out from a distant shore. I wish for you to see the sun shining so brightly

> on the day you first see the brown sea of heather! Those churches and houses, those heights and pyramids of peat would appear floating so much in the air, transforming their dark silhouettes every moment, and taking on the appearance of human beings, animals, trees, mountains and anything else the imagination can invent (Wyss-Neel 1971, p. 26)[3].

As fall approached, brown took on the violet hues of flowering heather and the blue of bluebells.

"The moor, certainly, you'd hardly believe it. But come and have a look for yourself! The heather forms a sumptuous carpet, flowers abound as far as the eye can see" (Wyss-Neel 1971, p. 5), wrote Denmark's most famous writer, Hans Christian Andersen.

In the late 19th century, Pontoppidan went furthest in the Edenic description of a pastoral scene:

> On a calm summer's evening, as the sun fades and casts like a golden glaze over every little pool of water, as the chubby-cheeked girls sing along the meadow paths, their strong shoulders laden with the yoke from which the milk jugs hang, when the red-haired boys waddle out of the villages on heavy horses, letting their hooves dangle to the tip of their big toe, when the marshes begin to bubble and the meadows to stretch their veils, it's then that you can truly believe you've been transported to the land of Cocagne, where everything exudes peace and endless happiness (Paul-Dubois 1909, p. 659; Pontoppidan 1926).

Long after he had traveled to latitudes ranging from the Arctic to the tropics, Eugenius would always express – in a more sober style – his visceral attachment to the landscapes and vegetation of his native land, to which he too found every attraction.

2.3. A demanding school

Eugenius attended one of the country's oldest "cathedral schools" (*Katedralskolen*), after Roskilde and Viborg. It was founded in 1145 on the site of

---

3 Texts by Blicher quoted from the complete works (*Samlede Skrifter*) in 33 volumes, published by the Danish Society for Language and Literature. Following Blicher's example, some of the most important Danish writers praise their native country: Drachmann, Knudsen, Skjoldborg, Aakjaer, Jensen.

an existing school. Originally, "cathedral schools" were created by bishops in association with their cathedrals, to provide the Church with an educated clergy. When Bishop Elias handed over the Ribe school to the cathedral chapter, there was a very fine Renaissance-style building, from the 14th century, with a monastic vocation, *Puggård*. In the early 18th century, new buildings came to rival the Copenhagen school.

Eugenius arrived just in time for the building's inauguration on September 5, 1856. In his speech, Rector August Bendtsen appealed to the Muses of the arts and sciences, to which the building was dedicated, for the good and benefit of the fatherland. The school was modern in its organization and facilities, yet rooted in a history symbolized by *Puggård* (Karsdal 1995).

A ministerial decree of May 13, 1850 opened up the curriculum to a new range of subjects, requiring more teachers, classrooms and places for assemblies. These included natural history, German, French, Latin and drawing. Eugenius developed a passion for the first subject and notable skills in the other four. The regulations specified the conditions a student needed to meet to be admitted to the university. Among other things, he had to be able to write in Latin without mistakes and speak it with a certain fluency. Eugenius was listed in the register for 1859, the year of his graduation. He was listed as an "illustrious alumni".

The key words of the "Disciplinary Regulations of the Cathedral School for Ribe Disciples" (*Disciplinariske Bestemmelser for Ribe Kathedralskoles Disciple*, 1848) were decency and dignified behavior. These recommendations also applied to the outside world. Everywhere and at all times, "students" had to behave in a moral manner. Rector Bendtsen (appointed in 1845) added a handwritten note to the regulations, forbidding students to walk with a cane (Paludan 1995). A dispensable accessory for the elegant man, this sign of social belonging was little appreciated by a Protestant who insisted on transmitting moral values based on humility. Day or evening canes, for the city, the country or travel, this panoply was outlawed. There was no question of Ribe's high school students strolling through the night sporting a distinguished cane, as Kierkegaard used to wear on the streets of Copenhagen.

Whereas in the past students thought only of having fun and drinking, now they had to take part in the cultural and associative life led by the *Ribe Byes forenede Klub-selskab* (Ribe United Club Society). A significant proportion of students came from the upper classes of society. They attended a school where they were expected to hold their own without ostentation and prepare themselves for a high social position.

Laurids Johannes Koch, a former pupil who left school in 1892 and went on to become a pastor and writer, reported that corporal punishment was customary, but proportioned to the offense committed and in a moderate manner. Floggings with soft canes or "nuts" (punches to the head) did not seem to be experienced as ill-treatment. In fact, when the choice was given by the teacher, preferring a slap to a reprimand was a kind of code of honor on the part of the pupils. Koch pointed out, as an exception, that one teacher earned respect "almost without hitting" (Koch 1945, p. 153). The teaching staff was made up of competent and respected teachers, such as Peter Trugaard, listed in Ribe in 1880, a fluent German speaker and specialist in French literature, and a believer in the "nut".

Figure 2.2. *Eugenius with other Ribe high school students (Fogh 1915, p. 7)*

It was a demanding school, requiring great self-discipline. The former student declared that he acquired a solid body of knowledge, habits of work and reflection, and classical training in Nordic and Danish literature, with instruction in Old Norse (the medieval Scandinavian language), Greek and Latin. His only regret was that the school's remoteness cut the students off from the movements that were driving the capital's youth at the time. When Koch went to school, after the Second Duchy War, Denmark was cut off from its southern territories. Between the two Duchy Wars, the situation was different. The Duchies of Schleswig, Holstein and Saxe-Lauenburg were still attached in personal union to the King of Denmark. This was Eugenius' situation, and he seemed to have enjoyed his high school years.

We see him posing with four comrades. They are all in uniform: a black jacket open over a buttoned vest, a white shirt and a bow tie. One of them, Pedersen, was for a time the editor of *Skolens Satyr*, the school newspaper. Eugenius became editor-in-chief. The authors published poems and prose essays of a satirical nature, dealing with school issues and teachers. They belonged to a group, the *Ribe Discipelforening* (association of Ribe students), which brought together "the best comrades of the 7th (A + B) and 6th class" (Fogh 1915, p. 7). Each meeting was "rather exclusive and secret, since the sons of civil servants and teachers, or pupils staying with teachers, were not allowed to take part". This is why a certain Nielsen, who appears in the photo, did not take part. They met every Saturday evening at the participants' homes, including, in 1857–1859, in Pedersen's shared dormitory at the Pontoppidan grocery store in Sønderportsgade (a Ribe street not far from the high school)[4]. Eugenius described this comrade, a future priest at Ribe Cathedral, as a rather cold person.

The group formed by the "disciples of Ribe" therefore had its own rules of admission and operation, and its own organ of distribution. Unfortunately, the group's successors were not up to the task, and their journal fell into ridicule. Be that as it may, the association disappeared in 1861 (but was reborn along with others).

2.4. The political context

In 1859, his last year at Ribe, Eugenius played a role in *Soldaterløjer* (*Soldiers' Jokes*). It was a musical comedy by Hostrup, with songs recounting the story of a farmer and his daughter, forced to take in Prussian soldiers during the First Duchy War. The choice of this entertainment was obviously not innocent.

The play was revived in 1909, at a time when Denmark was living under the regime imposed after the Second Duchy War. In Eugenius' time, there were only male actors, some of whom had to take on female roles. This was no longer the case in the early 20th century, as evidenced by a photograph taken in 1909 in which a woman (presumably playing the role of the farmer's daughter) could be seen posing on the theater stage set up in the gymnasium, along with six other actors, two of whom were wearing military garb and another the farmer's smock. A few spectators were seated, and a woman was at the piano (Fogh 1915, p. 46).

4 Probably of Erik Spend Pontoppidan (1824–1863).

Paa Scenen: H. A. Lund. S. Pontoppidan. E. Thorsen. Chr. L. Christensen. V. Tuxen. A. Petersen og Frk. Christensen.
Foran Scenen. Bageste Række: A. Bech Hansen. J. Poulsen. K. Fogh. Fru Dohn. Kr. J. Kristensen og Aa. H. Andresen († 1910).
Forreste Række: Pedel Chr. Nielsen. L. Koch. Frk. Laura Bentzon. A. Finsen. Musikdirektør Poulsen. Fru Overlærer Dybdal og Organist Behrends Clausen.

Figure 2.3. *Photo taken after the performance of* Soldaterløjer *(Soldiers' Jokes) in 1909 (Fogh 1915, p. 46).*

After 1864, Ribe and the surrounding area still belonged to Denmark. A former pupil of the Ribe school recounts an anecdote which shows that visiting former compatriots or family could be a risky business. One morning, a group of high school students set out to cross the border. A long line of wagons rolled along the Obbekær road, less than 10 km east of Ribe. Nervous German customs officers discovered two bottles that looked suspicious. The owners did not know the German translation of the contents, and the customs officers did not speak Danish. Two boys were taken to the station by two armed men who threatened them. A few hours later, they joined their comrades at an inn, where the latter were waiting for them. The heroes of the day arrived triumphant with their two bottles, which contained mead, a traditional drink known as "nectar of the gods".

The students felt like Danes facing a German front, although the teachers at Ribe were very neutral and never officially alluded to the situation. The majority of pupils were probably left-wing, and the teachers probably all right-wing, commented

Koch (1945, p. 148). During their time at Ribe, the students strengthened their national feeling and their resentment of their warlike neighbor.

It was also at Ribe that a teacher played an important role in Eugenius' further education. Carl Emil Nørager d'Origny, a native of Copenhagen, had been a Professor of Natural History at the Cathedral School since 1850. He was able to reinforce Eugenius' early interest in botany and develop his sense of observation of nature. In particular, he explored the meadows south of Ribe and collected plants. The high school student acquired a scientific background and developed study methods and discipline by frequenting the library, natural history collections and science laboratory.

Sixty years later, the former "student of Ribe", now a famous botanist, returned to his high school to celebrate a birthday of sorts. He was greeted by the principal at the start of the school year after the summer vacation.

3

An Interrupted University Course

3.1. Academic training

The Ribe diploma opened the doors to the University of Copenhagen. From 1859 onwards, Eugenius received a university education in botany, botanical geography and zoology, at a key moment in the history of science and his personal career.

He was not satisfied with Professor Ørsted's botany lessons, which he found boring. The student did not like wasting time. On the strength of his personal training and experience as a naturalist in Ribe, he decided to work on his own. Nevertheless, Ørsted left behind important publications in zoological and botanical systematics, taking into account the geological and physical conditions of animal and plant life. He was an innovator in marine biology, notably identifying food chains. He studied the distribution of plants and animals according to their habitats in relation to water depth, which he linked to light. He theorized that their distribution depended on the penetration and refraction of light (Fox Maule 2011a)[1]. Trips to Central America (Costa Rica, Nicaragua, the Caribbean) in the years 1845–1848 brought him back to botany.

His scientific and philosophical references were clearly identified. He drew on the work of the Dane Schouw, an important milestone in the history of ecological botanical geography. Ørsted's studies on plant structure were also influenced by Goethe's theory of metamorphosis and Schelling's *Naturphilosophie*. The former is based on the principle that the development of plants can just as easily lead to the reduction of organs as to greater perfection. In flowering plants, a single, changing organ can go from seed to leaf, sepal and petal, carpel, fruit, then back to seed, according to a rhythm of contraction and expansion. Progressive or ascendant

---

1 His theory about the link between algae color and the light spectrum turned out to be wrong, but he drew attention to the importance of this factor.

metamorphosis and descending or retrograde metamorphosis are explained by a homology in the structure of vegetative and floral organs, referring to a vital, original plant (*Urpflanze*). The same organs therefore serve different purposes and take on highly modified forms (Ørsted 1868; Guédès 1969). As an alternative to the mechanistic vision of modern experimental science inspired by Galileo and Newton, *Naturphilosophie* offers a metaphysical explanation for scientific discoveries, particularly in the fields of physics and biology, blurring the boundary between nature and spirit. It is "a romantic science of nature" rooted in the German idealism represented by Kant (Renault 2002, p. 53).

At Danish universities, certain decisions regarding the allocation of positions were influenced by these positions, which became political. Danish nationalists resisted German hegemony, both culturally and philosophically, especially in the context of the Duchy Wars.

When Eugenius received the teaching he was so unenthusiastic about, Ørsted was in the process of paving the way on which he himself would embark during his stay in Brazil. It is likely that Ørsted referred in his lectures to his "*Centralamericas gesneraceer et systematisk, plantegeographisk Bidrag til Centralamericas Flora*" (A systematic and phytogeographical contribution to the flora of Central America) published in 1858 – available for students to consult in the library – and alluded to his travels in Costa Rica and Nicaragua. He wrote the foreword to the first issue on July 1, 1863 (Ørsted 1858, 1863). This is all the more interesting given that Eugenius had the same twofold scientific intention as Ørsted: to explore flora both systematically and from the point of view of botanical geography. It is possible that he did not fully appreciate the contributions of the traveling naturalist until he knew that he too was going to spend time in a tropical region.

Ørsted, professor at the University of Copenhagen since 1821, director of the botanical garden since 1841, died in 1872. His succession gave rise to considerable tension, at a time when Eugenius could lay claim to his vacant post.

On the contrary, Eugenius declared that he followed with interest Vaupell's lectures on Danish forests. Vaupell was a professor (lecturer) at the University of Copenhagen[2]. Severely wounded during the First Duchy War, he had to have his left arm amputated to save him from gangrene. His health remained fragile due to the tuberculosis that affected him. His first thesis put him on the fringes of the academic world. When evaluated by the Royal Academy of Sciences committee, it was discredited. The reasons for this rejection are complex.

---

2 The spelling Kjøbenhavn (sometimes *Kjöbenhavn*) was officially replaced in 1889 by København, adopted by the town council in 1906, but the various studies cited in this book still used both forms well into the 1930s.

Vaupell, following in the footsteps of Professor Japetus Steenstrup, analyzed macrofossils from peat deposits and showed that during the Holocene, when forest communities were developing, birch was a pioneer tree, then came the pine, then the oak and finally the beech, indicating a forest stage tier in equilibrium with environmental conditions. But Steenstrup's theory of climatic control of these different phases of post-glacial forest succession was challenged by Vaupell, who emphasized soil-related factors. His work was published in Danish and German, and his conclusions were widely accepted in Germany and France, and then, at the end of the 19th century, by the founders of North American ecology. Vaupell's studies are now recognized as pioneering in Quaternary geoscience and in the study of vegetation successions (Vaupell 1851, 1858; Nielsen and Helama 2012).

The rejection of Vaupell's thesis was not based on scientific arguments alone. The influential Professor Steenstrup, considered one of the fathers of Holocene paleoecology, of whom Vaupell was a student, did not accept his pupil's conclusions, which differed from his own. But there were also power struggles within the evaluation committee, resulting in quarrels over the attributions of posts. His second thesis, defended in 1859, tackled a completely different subject: the reproduction and fertilization of a freshwater alga (*Oedogonium*). He obtained his doctorate and taught at the University of Copenhagen until his death in September 1862, at the age of 40. It is fair to say that Eugenius was able to benefit from his teaching in extremis.

Vaupell introduced the student to the study of vegetative reproduction and plant anatomy, using the microscope, as confirmed by Signe Prytz (1984), who collected letters from her grandfather, Eugenius Warming.

Eugenius was also taught zoology by Steenstrup. A popular lecturer, editor of scientific journals, politician, academic and member of the board of the Zoological Museum and the influential Carlsberg Foundation, he played a crucial role in the development of natural history in Denmark. As a professor of zoology from 1845, the university did not yet offer an academic degree in natural history. Thanks to him, it did in 1848. Two years later, he established a faculty of natural sciences, followed by a new zoological museum in 1870.

When Eugenius arrived in Copenhagen, he found a university offering complete academic training in natural history.

3.2. Debating evolutionary theories

Steenstrup discussed the theory of evolution with his students and corresponded with Darwin. As a New Year's gift, he received a copy of the first edition of

On the Origin of Species on January 1, 1860, with the author's compliments. However, as Darwin regretted, Steenstrup did not subscribe to his respected friend's theory.

Darwin never came to Denmark. However, before the publication of *On the Origin of Species*, he was well known in Danish scientific circles. In addition to Steenstrup, he corresponded with geology professor Forchhammer, who died in 1865. He was another of Eugenius' teachers. His thesis on the geology of the Faroe Islands was influenced by the *Principles of Geology* by Lyell, whom he met in 1834. Lyell showed that the Earth's history is made up of gradual changes over long geological periods. We know that this work had a major impact on the young Darwin during his round-the-world voyage on the *Beagle*.

Most of Darwin's books can be consulted at the Royal Library and the University Library – a new building was inaugurated in 1861 – by Danish students and academics, in the original English editions. The German translation of his work on coral reefs was available in several copies in 1844. Successive editions of his 1859 book were also registered, although the number of borrowers remained small during Eugenius' student years.

Danish naturalists generally regard Darwin as a respected English colleague. However, like Steenstrup, most did not agree with his theory of evolution and natural selection. It was discussed and even taught at the university, yet it was not considered to explain the history of life on Earth, but only as an interesting theory among others, proposed by one of the leading naturalists of the time.

Zoologist Reinhardt was an exception. He was of the same generation as Steenstrup; they knew each other well, as the latter had lived with him for a time. When Eugenius arrived at the university, Reinhardt was the equivalent of an assistant lecturer (1861–1865). According to his contemporaries, his lectures were well planned but boring. However, it can be assumed that Eugenius was assiduous and brilliant enough to be noticed and chosen by him to become secretary to the Danish naturalist Lund. Reinhardt made several trips to India, the Philippines, China, Japan and the Sandwich Islands. In Brazil, he became Lund's secretary. He spent several periods there between 1845 and 1856, when he taught zoology at the Polytechnic Institute, founded in 1829. In the meantime, thanks to his work and the collections submitted by Lund, he obtained the post of curator of terrestrial vertebrates at the Royal Museum of Natural History.

In Brazil, working with Lund on vertebrate bones discovered in caves not far from Lagoa Santa, he adopted positions critical of fixism and catastrophism, which postulates that species do not change over geological time, extinctions being due to local catastrophic events. When he read *On the Origin of Species*, he saw it as the answer to his questions. His writings from 1860–1863 testify to his adherence to

Darwin's evolutionary theory, at odds with his fellow zoologists, notably Steenstrup and his assistant Lütken. The latter, one of Eugenius' teachers, was also one of the first Danish scientists to discuss Darwin's theory with his students in Copenhagen, where he was a privat-docent between 1856 and 1862. This academic title permitted the holder to give lectures and conferences without being paid by the state. It is often a prerequisite for holding a professorship. Lütken communicated his views on Darwin in the popular science magazine *Tidsskrift for populaere Fremstillinger af Naturvidenskaben* (*Review of popular presentations of the natural sciences*), of which he was founder (1854) and co-publisher with geologist Fogh.

However, the general public was hardly involved in these debates, even though a brief notice about *On the Origins* was published in the popular *Illustreret Tidende* (*Illustrated Newspaper*) on February 5, 1860, under the not very evocative title: *Literatur*. Published from 1859 to 1924, the weekly magazine occupied a unique position in Denmark, offering literary, scientific, entertainment and international news. It was inspired by English and German equivalents.

Eugenius therefore entered university at the same time as Darwin's evolutionary theory. The discussions had to be sustained, between the majority of those who considered Darwin's evolutionism as one theory among others, which could be seriously discussed without adhering to it, and those – few – who were Darwinians from the 1860s (Funder 2013).

Eugenius formed his first opinion, close to that of another student, Johannes Steenstrup, son of Japetus. They already shared the same conservative political ideas, which they would defend when they themselves became professors at the University of Copenhagen. As we shall see, Eugenius' position had a profound influence on his ecological studies.

3.3. A harrowing crossing

One day in October 1862, while working in the former Botanical Museum of the Royal Palace at Charlottenborg, Reinhardt approached Eugenius and offered him a mission. The student made an important decision: to interrupt his studies to cross the Atlantic and spend two years in Brazil.

He was neither the first nor the last of Lund's secretaries. Before him, Norwegian painter and publisher Peter Andreas Brandt (sometimes spelled Brent) was an assistant and illustrator for Lund's publications, until his death in the fall of 1862. Brandt, who went bankrupt in the early 1830s, fled to South America, probably in search of a wealthy relative in Chile. But he could go no further than Brazil and ended up working for Lund (Sæther 2015, p. 33). Professor Reinhardt himself took on this secretarial task on several occasions. After Brandt's death, Lund asked Reinhardt, whose collaboration he had appreciated, to suggest a young Dane, preferably a botanist.

Eugenius embarked on February 17, 1863 for Leith (today a district of Edinburgh), a Scottish port where the French army had already landed in the 15th century, where he arrived 10 days later. He made the journey on the Danish brig *Marie Lehmann*, a light two-masted sailing ship bound for Rio de Janeiro via a trade route. He continued to keep his diary, in which he recounted the uncomfortable conditions in which he traveled. The young man, who had never sailed on the high seas, reported with horror on the storm damage to the sails. On board, he took advantage of his free time to fish for various animals, which he identified, to read, especially travelogues, and to learn Portuguese. Off the coast of Madeira, he noted on March 24 that the captain was in good spirits, which he said was exceptional. It seems that he took Eugenius on board at the last minute, under duress. He sometimes forbade him to fish, and refused to inform him of the latitude and longitude at which they were sailing. Eugenius complained of loneliness, lack of contact with the crew and the captain's pettiness towards him.

He found the time long, but his desire to arrive in Rio was not unmixed, he wrote on April 20. He was apprehensive about a stay in an unfamiliar city with unfamiliar people. However, when the coast of Brazil appeared in his binoculars on April 25, after a seven-week journey from Leith, he was relieved. The prospect of setting foot on this land was a promise of new-found freedom.

4

From Rio to Lagoa Santa: A New World

4.1. Free at last!

Arriving on April 27, 1863 in the wide Guanabara Bay, just as the rainy season was coming to an end, Eugenius discovered the capital of the Brazilian empire, already home to over 100,000 inhabitants.

The invasion of Portugal by Napoleon's troops led to the royal family and many nobles fleeing Lisbon for Rio de Janeiro. They expropriated part of the population, particularly those with the best housing. More than 600 houses with neoclassical facades were built between 1808 and 1818. The Italians, English and Chinese moved in, and the city underwent rapid development and exceptional modernization. The nobles opened the port of Rio to the international market. Previously, access to the bay had been restricted to Portuguese ships. Rio was the only city in the world to be home to a European empire outside Europe.

Eugenius' room at the Hotel Waltz offered him a magnificent view of the bay. He discovered orange, banana and other fruit trees. Apart from local produce, he noted that everything was expensive, having been imported from Europe. After more than two months on the road, he was finally able to resume his activities as a naturalist, visiting the botanical garden and, above all, exploring the flora of the surrounding area.

Claudel said of Rio that it was "the only great city [...] that has not succeeded in kicking nature out". Even today, the night air carries the scent of tropical rainforest descending from the wooded hills. It is a strange feeling to be in the heart of an urban area with a population of over 12 million, and to feel the same sensations as in Costa Rica, in the nebulous rainforest of the Monteverde Biological Reserve, over 1,300 m above sea level in the Tilarán mountain range.

A letter sent by Lund informed Eugenius that his future guide Manoel had left Lagoa Santa for Rio. In another, his maternal uncle informed him that his mother was unwell, but nothing serious it seemed.

Two weeks before setting off for Lagoa Santa, Eugenius sent a letter dated May 13, 1863. It accompanied a parcel containing specimens from several plant families (the legume family, euphorbiaceae, composaceae, etc.), intended for botanists in Copenhagen attached to the university and botanical garden. He made several such shipments during his stay.

4.2. A long journey

On May 28, 1863, he set off. His destination was 40 km north of Belo Horizonte, which became the official capital of Minas Gerais in 1897. He traveled 500 km in 42 days with a trailer. In the afternoons, he had time for excursions in the immediate vicinity. He and his guide had a horse at their disposal.

A map of his itinerary can be found in his *Lagoa Santa*, published in 1892. Eugenius wrote in his diary:

> Those leaving Rio de Janeiro have to cross the Serra do Mar [a mountainous formation in southern Brazil rising to an altitude of 2,000 m] to the north of Rio province. This route is one of the richest in natural beauty. [However], in many places, what was once forest is no longer a primeval virgin forest […], the axe and fire, in the service of harvesting, have produced clearings now covered with grasses and herbs, especially on the outskirts of the large farms or villages that appear here and there.

Once past the Serra da Mantiqueira (on the borders of Minas Gerais), "we entered a completely different nature, with new plants and new animals" (Ferri 1973, p. 15; Cavassan and Weiser 2020; Pereira de Lucena and Campolina de Sá Araújo 2020). The nature, the climate, the atmosphere, the sounds, everything was new:

> The drought in June and July is generally very severe in the countryside. The sun rises in a cloudless sky, with only a cloudy edge towards the mountains […]. There is almost no wind, but the air is pleasant and fresh; and in the fields, during the midday hours, the silence is so deep as to be oppressive, barely interrupted by the cry of the seriema.

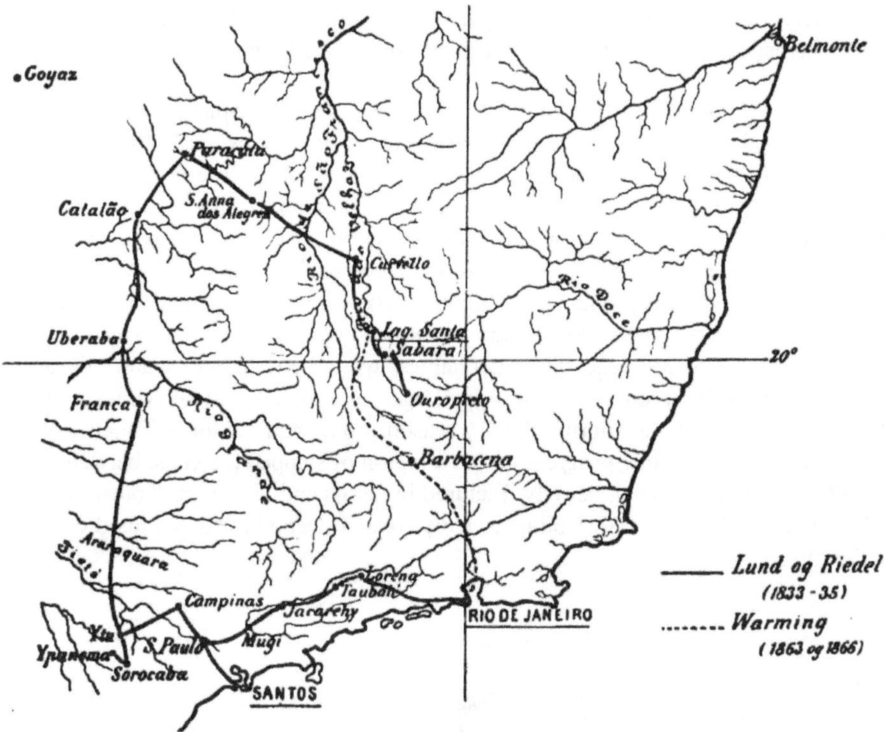

Figure 4.1. *Warming's (1863–1866) and Lund and Riedel's (1833–1835) routes from Rio de Janeiro (Warming 1892b, p. 115)*

Seriemas are the only current members of the Cariamidae family. These birds have a crane-like stature (90 cm) and brownish plumage. Their noisy calls are likened to yapping. They are often heard before they are seen. Eugenius appreciated these birds, which broke the silence of the quiet midday hours. He published a drawing of one (Warming 1868, p. 232). On the other hand, crossing paths with rattlesnakes gave him a fright.

Other observations and impressions are recorded by Eugenius in his diary:

> After two to three weeks, the path became steeper, my so much higher and higher. […] Once we reached the highest point, I turned and looked back. Before my eyes stretched an immense landscape, with ridges covered in pale green forests. When we reached the high plateaus, nature changed completely, in a strange way. In fact, the region was still full of fairly irregular valleys, but it was becoming increasingly deserted. There were only forests in the valleys, along the

> riverbanks, while the higher parts were covered with grasses and herbaceous plants, among which small trees spread out.

He was crossing and describing the *cerrado*, the neotropical savannah with trees that is characteristic of the Brazilian plateaus. It now extends over Paraguay and Bolivia (2 million km^2), and is a biodiversity hot spot (Moreira and Stehmann 2020, p. 144). For specialists, this formation is a biome (also referred to as a macro-ecosystem), i.e. an ecological unit that extends over a vast geographic area. The biome groups together ecosystems with comparable ecological conditions[1]. They are harsh, due in particular to the alternation between dry and wet seasons. From November to April, downpours are swallowed up by the crevices of the limestone rock and recharge the water tables, while torrents and waterfalls gush forth. Traveling between May and July, Eugenius crossed part of Minas Gerais, "Brazil's mountain state", during the dry season, from south to north. Together with his guide, they made slow progress across a region of stepped plateaus, between 600 and 900 m above sea level, heavily fractured by faults and diaclases, typical of karst reliefs (Kohler 1989, p. 63). Collapses have created deep chasms, sculpted caverns and grottos.

Figure 4.2. *The campo cerrado, photograph (1864) and drawing by Warming (1864; 1892b, p. 34, p. 62)*

Eugenius took photographs that clearly show the inextricable aspect of the vegetation, where it became denser. One of them, dated 1864, is entitled: "View of the *campo cerrado* at Lagoa Santa". His sketches perfectly capture the woody character of the shrubs and the leathery appearance of the leaves. One of them,

1 The term and its definition are thanks to Clements (1916) and Shelford (1931).

completed at the end of the dry season, shows ghostly shrubs, twisted and stripped of their foliage[2] (Klein 2000, pp. 50–51, p. 53 *ff*; Holten et al. 2004, pp. 219–220).

"Three or four days' drive from Lagoa Santa, the vegetation has become more abundant [...]. The trees are straighter, taller and closer together, but the thick red clay that makes up the soil continues to be covered with tall and numerous herbaceous plants". This is how he discovered the sinkholes that are common in this area of highly karstified plateaus. As a botanist already attentive to the conditions in which plants grow, Eugenius noted the presence of clays (*terra rossa*, red earth). The physiognomy of vegetation changes, as water retention makes the soil more favorable to tree growth and herbaceous development. As a result, islands of humid forest, meadows and clearings can be found at the bottom of basins and valleys.

"From then on we walked, day after day, through this region. From time to time, at the bottom of the valleys, there was a farm where we spent the night. Bivouacs were offered in a few villages of some importance": J. de Fora (670 m above sea level, officially named in 1865), then in full expansion, Barbacena (in the Serra da Mantiqueira, 1,160 m above sea level), or Bom Fim, a camp for cattle herders.

His guide Manoel, a *fazendeiro* (farmer) who owned a herd of mules, helped them take advantage of sections of the newly opened *União e Indústria* (literally: unity and industry) road (1861) to facilitate the transport of coffee, transiting through J. de Fora (Cavalcanti 2019). Eugenius described his guide as accommodating and trustworthy, unlike his "dear ship's captain" (Prytz 1984, p. 25).

When Eugenius finally arrived in Lagoa Santa, he met a thin, grizzled man, moving with difficulty, in the shape of Lund. An impressive character for a young student, this scientist was known far beyond the borders of Denmark, and had been living for years on the fringes of the scientific community. He was relatively isolated in his house in Lagoa Santa, facing criticism and controversy for his bold hypotheses relating to his paleontological discoveries. But Eugenius was confident. He had been recommended by his teacher.

The relief of arriving in Lagoa Santa and the emotion of meeting a great scientist were overshadowed by the letter from his uncle. A few weeks after hearing the sad news, the young orphan wrote to him: "The death of my unforgettable dear mother has kept me too busy to write at length and in detail. It's obviously very nice what

2 Warming may have had access to some of Brandt's illustrations and drew inspiration from them, or even, according to one polemic, unduly claimed authorship. He is even said to have discredited Brandt, whose social background was inferior to his own.

the priests said at her funeral, but there's no one who, like me, knew her devoted and warm heart". These last words speak for themselves, such is the only son's attachment to his mother. As the only direct descendant, it was now up to him to settle the succession:

> In the days that followed, I was busy drawing up various legal documents relating to my mother's death, which, after being signed in the presence of the local authorities, were sent to the higher authorities in Sabará [county town some 30 km from Lagoa Santa] and finally to the Consul General in Rio. These were several procurations, drawn up according to a model sent by Han Ollgaard to the authorized representatives in Denmark, who were to deal with my affairs.

Due to mail delivery conditions, these issues took many weeks to resolve.

4.3. Eugenius' employer

4.3.1. *A land of opportunity*

Born in Copenhagen on June 14, 1801 into a wealthy merchant family, Lund studied medicine at the University of Copenhagen. He made his first visit to Brazil in 1825, settling in a fishing village. Afflicted by the beginnings of tuberculosis, he saw in the Brazilian climate the promise of a cure, by escaping the mists and cold of the Scandinavian winter. His father and two brothers had already been taken by the same disease.

An all-round naturalist, he spent three years observing and collecting plants, birds and insects around Rio. He made a few excursions to Niterói (on Guanabara Bay, opposite Rio) and longer stays near Nova Friburgo, in the mountains (not far from pico da Caledônia, altitude 2,250 m) (Lund 1845–1849; Darwin 1859, p. 339; Winge and Winge 1888; Rivet 1908; Stangerup 1983; Luna Filho 2007; Mackenthun 2016)[3].

Back on the Old Continent, he published a dissertation in German, which quickly won him international renown. The "standard German" used in texts for populations using different Germanic dialects became, after Latin, the vehicle for scientific communication in 19th-century Europe. Lund defended his thesis in 1829 at the University of Kiel, went to Hamburg, traveled to Italy – partly for health reasons – settled in Paris, where he was deeply influenced by the work of the great paleontologist Georges Cuvier, holder of the chair of comparative anatomy at the

3 Biographical and scientific information on Lund is taken from several authors.

Muséum national d'histoire naturelle and a member of the prestigious Royal Society of London.

Lund adhered to the catastrophist theory defended by Cuvier, according to which the extinction of species was caused by natural catastrophes (floods, earthquakes) – "revolutions of the surface of the globe" – in certain regions of the world (Cuvier 1830)[4]. Repopulation was thought to have taken place through new creations or migrations. In other words, disappearances and appearances were the result of local crises. This theory was in line with the fixist paradigm defended by Cuvier, which stated that created species do not transform (Buican and Grimoult 2011, Chapter 4).

Figure 4.3. *View of Lagoa Santa. Foreground:* Sagittaria lagoensis *(synonym:* S. rhombifolia*) (Warming 1892b, p. 189)*

Lund returned to Brazil in 1833 and spent two years in Rio, devoting most of his time to botany. Still in poor health, he eventually settled in Lagoa Santa. Was he attracted by the waters of the lake, known locally since the 18th century for their medicinal properties? Felipe Rodrigues, founder of Lagoa Santa in 1733, is said to have relieved his aching legs in the waters of the lake. From then on, pilgrims went there to seek a remedy for their ailments. The first chapel was built in 1749, followed by a church in 1819, Nossa Senhora dos Remedios (Our Lady of Remedies), also known as Nossa Senhora da Saùde (Our Lady of Health). The parish of Lagoa Santa was created in 1823. There were other scientific travelers, such as the Swiss-American zoologist Louis Agassiz, whose health deteriorated in

4 Georges Cuvier excluded the human species from his theory.

the 1860s, or the Swedish botanist Anders Fredrik Regnell, who settled in Minas Gerais in 1840 after suffering from a serious lung disease, and remained there until his death in 1884.

Since the 16th century, many European travelers have been drawn to South America (Laissus 1995; Kury 2001). For the French, it is a land of choice. For instance: Thevet (Brazil), Bouguer (Peru), Dombey (Peru, Chile), Bonpland (Argentina), the traveling companion of the great naturalist von Humboldt, d'Orbigny (Brazil, Argentina, Paraguay, Chile, Bolivia, Peru), Boussingault (New Granada), etc. were the source of major scientific contributions. The great voyages of exploration are also titles of glory. Publishing reports, holding exhibitions and giving lectures all contributed to building a career and enhancing the author's reputation. Von Humboldt was often cited by Warming for his work in botanical geography, after his great voyage to "Equinoctial America" (1799–1804).

4.3.2. *A discovery*

The source of further torment for Lund was a discovery. To understand the Danish scientist's situation when Eugenius arrived in Lagoa Santa, we need to go back to the 1830s.

Clausen (or Claussen), from Copenhagen, was a scandalous character – naturalist, fossil collector and dealer. He emigrated to Brazil following a series of frauds, and from there became a soldier, a spy, a merchant and then the owner of a farm in Curvello, a few days' walk from Lagoa Santa. Lund had a chance encounter with Claussen while touring the region with the German botanist and explorer Riedel, who worked with him from 1833 to 1835. In 1836, Riedel became the first person from overseas to hold a permanent position at Rio's National Museum.

They were interested in saltpeter from limestone caves. Lund learned from Claussen that the Maquiné cave (named after its discoverer) in Cordisburgo was home to large bones attributed by locals to gigantic prehistoric men. Located at a depth of over 15 m, the excavation was over 650 m long. In the 1880s, a tiny palpigrade endemic to Minas Gerais was found, *Eukoenenia maquinensis* (of the Arachnida class). Today, the cave is visited for its impressive dimensions and the beauty of its stalactites. In front of the bones discovered in 1834, Lund is said to have exclaimed: "Never have my eyes seen anything so beautiful and magnificent in the realms of nature and art".

In the 19th century, scientists no longer believed in the myth of an ancient race of giants having populated the Earth, which was much talked about during the Enlightenment. The local inhabitants were undoubtedly impressed by the bones of large mammals (camelids, deer, sloths and giant armadillos), which they mistook for those of a race of giants. The Hominid fossil bones that Lund studied for years correspond to individuals ranging in size from 1.43 m to 1.53 m for adults.

Between 1834 and 1843, he doggedly excavated over 200 caves. By February 1838, he had reached 88 caves, rising to 106 by September. He sometimes had to rope up to reach some of them. In front of the Maquiné cave, he even built a mud hut for protection and, no doubt, to spend the night in. His health problems did not yet affect him too much.

Ultimately, only six caves contained fossil Hominid bones, including the Lapihnas cave, 90 km north of Lagoa Santa. Formed 600 million years ago, it is now part of Sumidouro State Park, created in 1980. Considered the father of Brazilian paleontology, Lund has his own museum there.

4.3.3. *Controversies*

When Lund entered the debate on the origin of mankind, fossil bones were no longer regarded as "fancies of nature" or "relics of the Flood". Paleontology – the term was coined in 1822 by the French zoologist Ducrotay de Blainville – is based on osteological collections studied within the framework of comparative anatomy developed by Cuvier.

Lund wrote in a long letter dated March 28, 1844, sent to the Danish archaeologist Rafn and published by the Royal Nordic Society of Antiquaries, founded in Copenhagen in 1825, that he had discovered a Hominid in a vast cavern "on the southern shore of a lake called Lagoa do Sumidouro", 20 km north of Lagoa Santa, and that the bones "belonged to at least thirty individuals of different ages, from newborns to decrepit old men". In his opinion, they "belong to a very remote period". However, he did not go into chronology. Proposals concerning the supposed age of the Earth abounded. For the 17th century theologian Ussher, the Earth was created on the night before October 23, 4004 BCE. In the 18th century, the naturalist Leclerc, Comte de Buffon, experimentally determined the age of the Earth, finding 74,047 years. Darwin put forward a figure of 100 million years around 1860, later returning to a range of 20,000–40,000 years.

Lund submitted the collected bones to the scientific community, first by sending them to the King of Denmark, who presided over the Royal Nordic Society of Antiquaries. He also sent a small collection of 36 issues to the same society. Among its founding members were the Emperor of Brazil (Dom Pedro II), the King of Greece (Othon I) and leading figures from Russia, England, China, Spain and Italy.

The age of the bones discovered was fiercely discussed, debated and doubted. In a volume devoted to the human and animal bones excavated by Lund and preserved in the paleontology department of the Zoological Museum at the University of Copenhagen, anthropologist Sören Hansen lamented: "Never were humans associated with animal bones in such a way that one could deduce an absolutely certain contemporaneity with a tertiary or quaternary fauna". Hansen himself carried out an in-depth study based on 15 skulls. He then raised an issue that was hotly debated in the 19th century, that of the existence of antediluvian man. "Various interpretations were to fuel debates on the existence of Man and the existence of Tertiary Man". The Tertiary corresponds to what we now call the Quaternary (Hansen 1888, p. 35; Elizondo 1990; Grandchamp 2009, p. 37; Patou-Mathis 2012, p. 10).

Lund's reconstruction of the geological context in which the bones were found is based on paleontological, geological and ethological arguments, i.e. those relating to the behavior of the animals and therefore to the way in which they would have arrived in the cave. For example, he believed that the hinds wandered into the caves to lick the saltpeter and were unable to find their way out.

His arguments show that he has incorporated certain elements of British geographer Charles Lyell's *Principles of Geology,* particularly when he reconstructs the regional history with its gradual changes over long periods: subsidence, uplift of limestone strata that once formed a plain and now a plateau, etc. However, Lund had effectively neglected to carry out stratigraphic surveys; it was difficult to convince his contemporaries that the human remains were associated with a fossil fauna (Lyell 1830–1833; Gohau 1990, p. 156 *ff*; Julien 2000, p. 775; Da Gloria et al. 2017). Even though he admits, at the end of his 1844 letter already quoted – taking the form of an argumentative demonstration – that his opinion was not yet fully founded, he put forward quite firmly that of the "total reversal of the chronological relationship which has hitherto been established between the two races of which we speak". These were the Mongolian and the American races.

The question was posed in these terms at the beginning of the 20th century by Henry Vignaud, an American scholar, historian and diplomat: "Would Man have appeared originally on different points of the globe, and would we have to see, in the one found in America, a human type not generically related to any other?" His answer was clear: the fossils discovered there were more recent than those of the Old World, and the peoples of the Americas had long remained in a primitive state, whereas those of the Old World had developed civilizations dating back thousands of years. The author stated "categorically [...] that if man is the result of the evolution of an anterior form, we cannot place this evolution in America" (Vignaud 1922, pp. 18, p. 22).

This was the crux of the problem, which went far beyond the scope of scientific argument. The idea that Europe could have been colonized secondarily thanks to the immigration of the first men to appear in the Americas was unacceptable and even unthinkable. To admit that the inhabitants of the New World could have preceded those of the Old World would require a veritable Copernican revolution, a "total reversal of the chronological relationship established up to now".

Foucault wrote:

> Each society has its own regime of truth, its "general politics" of truth: that is, the types of discourse it accepts and makes function as true; the mechanisms and instances that enable true or false statements to be distinguished, the way in which one or the other is sanctioned; the techniques and procedures that are valued for obtaining truth; the status of those who have the task of saying what functions as true (Foucault 2010, p. 112).

This "reversal" evoked by Lund clashed with the 19th century regime of truth, which supposedly allowed true knowledge about the origins of humanity to be socially and institutionally grounded.

A follower of Cuvier in his youth, Lund distanced himself from his master's ideas when he realized that his own observations were not consistent with the catastrophist theory. We have seen that he was influenced by Lyell's *Principles of Geology*, which give a uniformitarian vision of Earth history, opposed to catastrophism. Having noted the difficulty of establishing clear boundaries between an extinct species and its successor, Lund applied himself to looking for transitions in time. However, he undoubtedly lacked intellectual audacity in not proposing a genuine alternative to the paradigm that framed his early work (Luna Filho 2007, pp. 35, p. 141–142).

It does not seem that Darwin and Lund met, even though they were in Rio between early April and early July 1832. In any case, Darwin would later regard his paleontological discoveries with interest. In Chapter X of *On the Origin of Species* (1859), on the "geological succession of organic beings", he writes:

> Professor Owen has shown in the most striking manner that most of the fossil mammals, buried there in such numbers [various localities in La Plata], are related to South American types. This relationship is even more clearly seen in the wonderful collection of fossil bones

> made by MM. Lund and Clausen, in the caves of Brazil (Darwin 1859, p. 339).

Zoologist Owen worked extensively on Darwin's rich collection of South American fossils.

Lund was going through an intellectual and epistemological crisis. He abandoned paleontology to return to botany, to which he had devoted himself during his first stay in Brazil. Did botany, then considered "the most amiable of the sciences", bring him some peace?

He died before he knew he had discovered the "Lagoa Santa Man", believed to be South America's oldest representative. Professional excavations carried out in 1956 by Hurt and Blasi date the fossils to between 10,000 and 9,000 years ago (early Holocene). The skeleton of "Luzia", discovered at Lagoa Santa in 1974, has been dated to 11,500 years ago.

4.3.4. *Relative isolation*

De Luna Filho, a historian of science at the University of São Paulo, questions Lund's claims that he lacked the means to continue his paleontological work. In fact, his family funded all of his research for 35 years. A romantic hagiography – maintained by Lund himself – is said to have cultivated the image of the isolated researcher, far from home, abandoned by all, feeding the myth of a happy childhood, followed by poor health and a frail old age. It has also been suggested that Lund did not wish to return to politically unstable Scandinavia.

His scientific proposals, judged inadmissible by many of his peers, and the polite obstinacy he displayed in his letters, also built up the image, always ambivalent, of the cursed scientist (de Luna Filho 2007, pp. 285–286, pp. 298–299)[5].

By the time Eugenius arrived in Lagoa Santa, the weakened man had long since stopped visiting caves and karst caverns. His explorations of unhealthy and dangerous excavations, which had taken their toll on his health, had ceased from 1844 onwards. Afterwards, he spent much of his time corresponding, notably with the curators of his collections in Copenhagen, and studying his fossils in a barn located behind his house. Less isolated than has been claimed, he received visitors, scientists and illustrious figures such as the zoologist Hermann Burmeister, who explored Minas Gerais in 1850–1852, and the explorer Richard Francis Burton in

5 See in particular Holten, "To recall to life an unexpected treasury – Letters by Peter Lund" (undated), quoted by de Luna Filho.

the 1860s and, of course, Reinhardt on three separate trips. Geologist and paleontologist Orestes Hawley St. John, a member of Agassiz's Brazilian expeditions, confided his disappointment at finding so little in the caves excavated by Lund (Agassiz 1869).

Two years after Warming's return to Denmark, Riedel accompanied Luis Augusto, Duke of Saxony, son-in-law of Emperor Pedro II, on an expedition to Brazil, during which the group visited Lund. Among the photographs in Thereza Christina Maria's collection (21,742 photographs) collected by the Emperor, and bequeathed to the National Library of Brazil, are those of Lund's house.

Lund had no complaints about Warming – quite the contrary. In his later publications, Warming always paid him a heartfelt tribute. For example, in the introduction of his *Lagoa Santa* of 1892, he begins his text with acknowledgments and ends it by insisting that the scientific contributions in regional geology and paleontology are due to Professor Lund. The Central Library of the University of Copenhagen's Botanical Museum preserves 20 letters written by Lund to Warming, as well as his diary, which he gave him as a gift. Warming was also responsible for the only photographs of an elderly Lund.

The old scholar's life, both physically and morally, did not improve after Warming's departure. In a letter dated July 7, 1877, he complained of financial difficulties that prevented him from publishing. He described the violence of the rains, which destroyed buildings and bean crops, followed by extreme drought. He said he was weakened by a recent crisis that had further affected his lungs, to the point where he could no longer properly tend his garden. He also informed him about life in Lagoa Santa: deaths and new arrivals. He greeted the announcement of the railroad's imminent arrival in Barbacena, and mentioned some of the species Warming had sent him. The last letter sent to Warming is dated March 28, 1880. Warming replied on April 4, and Lund died on May 25. The villagers were saddened by the death of the man they called Doctor Lund, consulted by the sick. For his funeral, he asked for music and forbade tears (Dos Santos 1923).

He had long since named his brothers Henrik Ferdinand and Johan Christian as his sole heirs. Confident of dying young, he drew up his will in 1825. However, in 1875, he asked them to look after his adopted son, Nereo Cecilio dos Santos, his wife and his two daughters, by paying them each 300 rigsdalers. He also bequeathed Nereo 30,000 rigsdalers (this currency was used until 1875 in Denmark, the West Indies, Norway and Greenland). Nereo received a literary, scientific and musical education from his adoptive father, learned Portuguese and had a reasonable command of several languages (French, Danish, English), with a preference for French.

As a Protestant, Lund could not be buried in the small Catholic cemetery of Lagoa Santa. He planned to be buried in a plot acquired for this purpose, alongside his friends and collaborators Brandt, Behrens (who remained in his service for only a short time, due to his drinking habit) and Müller, in the shade of a typical *cerrado* tree, the pequi (*Caryocar brasiliense*). Today, his grave is practically in the city center. A monument dedicated to Lund and Warming was erected on the initiative of the Minas Gerais Academy of Letters.

4.4. The economy of collecting and collections

4.4.1. *"Explore thoroughly and travel less" (de Candolle 1855, p. 1348)*

In addition to his secretariat, Lund entrusted Eugenius with the task of continuing the botanical work he himself had begun and abandoned. He no longer had the physical resources to continue them. Eugenius was to be his eyes and legs. Lund regarded him as a very promising naturalist and a hard worker. However, Eugenius regretted that his service left him only five to six hours a day for his botanical errands, which were usually carried out alone and in the hottest part of the day, typically after 9 a.m. He took his meals with Lund, and in the evenings, he had to make himself available. He stayed in a small house built in the garden. He found that his employer's health, both physically and mentally, was relatively good, although his eyesight was impaired and he had memory problems.

The young naturalist's strategy was clear. He claimed to have followed the advice of Swiss botanist Augustin Pyramus de Candolle: "Explore thoroughly and travel less". This approach evoked the choice made by many French provincial naturalists, who in the 19th century relentlessly explored their "little local homeland", often limited by administrative boundaries (departmental, cantonal or even communal), which they prided themselves on knowing better than anyone else.

In Eugenius' case, his territory was far from his beloved homeland, and everything there was new, even if he claimed to be following in the footsteps of Lund and Reinhardt. While he was constrained by the need to ensure his service to Lund, he was also constrained by the length of his stay. His reference to de Candolle was a form of secondary rationalization, as his time was short and his task immense. He had originally planned to stay for two years, but was persuaded by Lund to extend his stay. After arriving in Lagoa Santa on July 8, 1863, he said farewell on April 24, 1866, but not before trying to persuade Lund to come out of his isolation and return to Europe.

He traveled the region almost always on foot, sometimes on horseback when the distances were too long and the paths allowed it (Lapa du Bahu, Sumidouro and its lake, Lappinha, Fazendaen Boa Vista). He was interested in the cliffs of Sumidouro, where there is a lake in which, he wrote, birds come to fish peacefully. He observed herons, colorful birds such as the *Parra jacana*, whose long toes enable it to move easily over floating plants, and the roseate spoonbill (*Platalea ajaja*), a wader with long legs and a spoon-shaped beak. About 30 km to the southeast, he prospected an area in the Serra da Piedade (Caeté). With hindsight, however, he felt that the trip, while botanically interesting, was a waste of time. After barely three years of excursions, he had meticulously combed an area of 170 km^2 – in a later publication he spoke of 150 km^2 (Warming 1899, p. 2) – with Rio das Velhas forming a boundary to the east, like the small Ribeirão da Mata river to the south.

Lund offered him his collections and a compilation of his notes: "observations on the vegetation of the interior plateaus of Brazil, particularly with regard to the history of plants" (Lund 1835). This work, undoubtedly his main contribution to botany, describes three types of plant formations: the forest, the *cerrado* with its shrubby vegetation of twisted, thorny trunks, and the fields dedicated to agricultural and pastoral activities.

Eugenius incorporated these data into two important publications, for the work they represent and for their contributions to the fields of floristics and botanical geography of the Lagoa Santa region. The first was the flora of central Brazil, *Symbolae ad floram Brasiliae centralis cognoscendam*, the first elements of which appeared between 1867 and 1869, the last in 1893, and a book on the geography of the plants of Lagoa Santa, *Lagoa Santa. Et Bidrag til den biologiske Plantegeografi*, published in 1892 (Warming 1867–1893, 1892b, 1893a; Ferri 1973)[6].

First of all, he had to confront unfamiliar environments, and manage sample collection, storage and preservation conditions for which he was ill-prepared.

4.4.2. *A difficult harvest to manage*

In December 1864, Eugenius wrote to Professor Lange: "With regard to the conservation [of plants] last year I had quite a few sad experiences which shattered my dreams". He mentions herbaria ravaged by insects, then, during the rainy season, by humidity (Prytz 1984, p. 29). Many years later, he expressed his regrets at length, and made no secret of his mistakes and blunders, those of a very young naturalist, which he asked readers to forgive. "What I regretted; the last year didn't bring me a similar return in scientific terms, the sacrifice was too great". He added, however:

6 The 1892 edition was translated and a foreword written in Portuguese by Löfgren in 1908. The 1973 edition is a facsimile of this translation, with text by Ferri.

> If I had thought from the outset of staying at Lagoa Santa for about three years, I probably would have taken things more calmly than I did, which would have been of great benefit to my botanical work, because I set myself the goal: to make a complete collection of the plants of the region, and as at the beginning of my stay I was completely overwhelmed by the richness of tropical nature, I doubted whether I would be able to achieve this goal in the two years planned, I failed, in a nervous zeal to do everything and organize my collections as well as I would have liked. […] Another circumstance that was also somewhat detrimental to the achievement of this goal, but certainly not to the detriment of my education as a whole, was that, for my knowledge, I thought it necessary to make long and detailed descriptions of living plants; I wasted a lot of time describing what could as well be seen on the dry plant as on the living one; I very much regretted that this time was not used to make even more excursions and collections. Finally, I had a few misadventures. One was that, at the end of the first rainy season, I discovered that a number of plants in my collections were moldy and had to be thrown away. I'm not sure I've replaced them all. The second is: I collected a lot of animals and plants in alcohol, but whereas in almost all cases I used glass containers for the zoological collections, I thought it was enough to utilize tin cans, made by the local schoolteacher and used for merchandise, for the plant collections. I overfilled each vase and didn't change the spirit [of wine, an alcohol obtained by distilling wine] often enough. The consequence was that on my return to Copenhagen, it turned out that the tins were largely infested with rust […], the whole thing had to be thrown away […]. I thought I should mention these circumstances, as they contributed to the fact that my floristic list cannot include everything that grows in this small area of a few square kilometers. Another circumstance, of course, is the infinite species wealth of tropical nature itself (Warming 1892b, p. 11).

In the end, he succeeded in bringing back a sizeable collection, as can be seen from the wealth of exchanges and donations. On the way back to Rio, he loaded his equipment and 14 boxes of his precious harvest onto mules. Not without emotion, he records having heard the cry of the seriema for the last time.

In Rio, he waited for a Danish schooner to take him back to Southampton for a reasonable price. He had spent a considerable amount of his budget on mules, horses and other expenses. He was affected by rheumatic pains and dizziness. A doctor consulted on the spot told him the causes were humidity and heat, too rich a diet and too much coffee. His family correspondence seems to show that he was bothered by rheumatism all his life. From 1910 onwards, he was most often disabled, to the point of refusing certain excursions.

On boarding, he declared to customs that his cargo was destined for the Copenhagen Museum. During the voyage, he fished animals in the Sargasso Sea for Professor Japetus Steenstrup.

On his return, Warming was quick to entrust parts of his herbarium to specialists all over Europe for study (Berlin, Paris, London, Geneva, Munich, Brussels, Hamburg, etc.). Among them were authors who contributed to the development of ecology, discussed below (Grisebach, Engler, Drude). For this, he drew on his 3,000 exsiccata (herbarium specimens of dried plants) and the 700 others from the collection donated by Lund. "Some [herbaria] survived the siege of Paris and the fires of the Municipality unscathed," wrote Warming (1892b, p. 12). Some botanists worked for more than 20 years; others died before finishing or gave up. Finally, he mentioned some 50 collaborators to his *Lagoa Santa*.

Warming did not wish to keep and store his collections. As early as 1868, he donated to the Copenhagen Botanical Garden a barrel containing fruit preserved in alcohol, a collection of wood samples, pieces of vine and dried fruit, with the proviso, he insisted, that it pay for the glass and alcohol and provide him with living plants in exchange[7]. In a letter dated January 27 1887, he wrote to Gray, a specialist in North American plants, with a crate of herbaria (including 90 Mexican species) and 190 Brazilian species collected by himself. In exchange, he asked him to send Arctic species from America and Greenland that were missing or rare, and plants from North America in general. After a trip to Greenland in 1884, he wanted to compare boreal and arctic flora. At the same time, his herbarium, enriched with other items, became the property of the University of Copenhagen. He was quick to share his collections with his peers and students. Many naturalists, on the other hand, bequeathed their collections in their wills, or their descendants donated them to an institution (museum, library, botanical garden, school, etc.). Some collections were dispersed among several amateurs, or even sold in shares, along with the equipment and works from the naturalist cabinets of the deceased.

Warming was not a collector. He was a man of the field and a researcher who needed a great deal of "material" to conduct his studies. He used this term several times in the letter to Gray, when he asks for specimens to work on.

Finally, despite the regrets he wished to share with his readers a quarter of a century after his return, his long and meticulous exploration of the Lagoa Santa region was very fruitful. He identified 2,600 species of vascular plants (with

7 Orchids, Bromeliads, Marantaceae, Iridaceae, *Peperomia*.

sap-conducting vessels), estimating that over 400 were new (he ultimately retained approximately 300 species) (Warming 1892b, p. 2; Moreira and Stehmann 2020, p. 145)[8]. He had dreamed of determining all the species in the area, which he estimated at at least 3,000, but ran out of time and had a few setbacks.

4.4.3. *Proof by object*

Solving the problem of preserving and transporting harvested samples is fundamental for naturalists. In botany, the discoverer indicates the location(s) as precisely as possible, describes the fresh plant – in particular its plant habit, color, etc. (characteristics that cannot be observed or are altered by herbarium storage) – the physiognomy and conditions of the site (soil, orientation, topography, humidity or dryness, etc.), and the distribution areas if possible. His peers checked the identification made by the collector.

Young Eugenius, inexperienced, ill-informed or ill-advised, and no doubt caught up in the imminence of his departure, was unable to procure and transport the appropriate equipment. He had to improvise. He was not the only one, and too many samples arrived at the museums in Europe in a disastrous state, which nevertheless published instructions for the packaging and transport of natural objects (Verlot 1865; Capus and Rochebrune 1879; 1883; Kuhlmann 1947; Schnell 1960; Allain 2000; Kury 2001, p. 129 *ff*). The loss of samples was dramatic. Until the advent of consumer macrophotography in the 1950s, followed by DNA barcoding and mobile applications that enable plant identification in the field, it was necessary to collect, for oneself and one's peers, in order to confirm the determination of each species.

Samples serve as evidence. They are passed from hand to hand to be observed at meetings of learned societies, sent to specialists in difficult taxa, collected in several specimens and placed in herbaria. Until the beginning of the 20th century, botanists did not hesitate to collect abundantly, for themselves and their colleagues. In fact, learned societies often instructed them to do so: amateurs "centurized" the rarest plants, as they were prized by collectors. They made up packets of 100 identical plants, like so many Roman divisions. The seeds, if collected in the right season, were sown in some private or public gardens, sometimes managed by local societies. Transplants were carried out when sufficiently fresh whole plants were available. At this point, discussions began on the conditions favorable to their successful cultivation, which gave rise to experiments for the acclimatization or naturalization of certain species.

8 Moreira and Stehmann listed 2,420 species of flowering plants. The revised list, combined with herbarium databases, gives 2,512 species.

4.4.4. *Equipment, techniques and innovations*

For a day's excursion, botanists have at their disposal special equipment: the green metal box familiar to naturalists is fitted with two lids closing a large compartment for plants, and a small one for seeds. Its straps make it easy and elegant to carry over the shoulder.

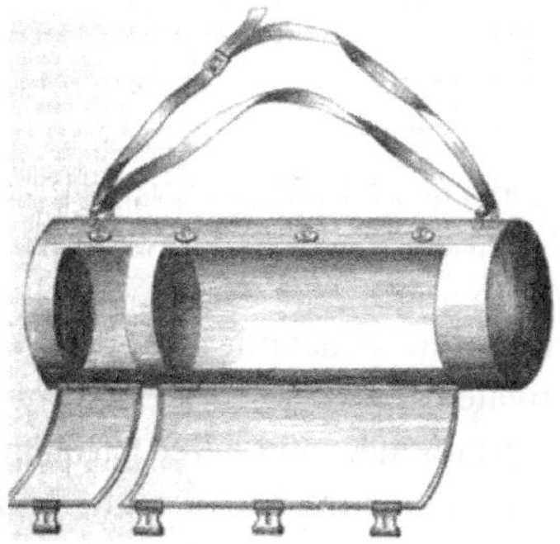

Figure 4.4. *Herbarium box (Verlot 1865, p. 38)*

Excursionists prefer a specially designed binder for transporting two-dimensional plants: a portable press. In France, as in the rest of Europe, specialized companies provide naturalists and arboriculturists with instruments. Pickaxes by Hacquin, Decaisne or Cosson (the latter combines a pickaxe and a hammer, useful for collecting saxicolous plants, pioneer species that grow in rock crevices), Deyrolle and Rivière crooks, pruning shears and, of course, a magnifying glass worn around the neck like a decoration, a sign of belonging to the group of "brothers in botany" (in the 19th century academic societies were dominated by men)[9] (Decaisne 1858–1862–1875; Cosson 1872).

In the tropics, harvesting, storage and transport are more perilous and the results are more uncertain than in temperate climates. Eugenius learned this the hard way. The collection already mentioned was partly different (the green box was of little use, given the size of most samples) and completed by a machete and a pole with a sling, which was a pruning shear attached to one end and operated by a cord. The weeder could be replaced by a sharp hook or even a fruit picker. The aim was to

9 Hacquin was a Parisian horticulturist and member of the *Société botanique de France*. Decaisne was a gardener at the *Muséum national d'histoire naturelle* in Paris. Cosson was a botanist. In 1831, the Deyrolle family (entomology, taxidermy, scientific and educational equipment) founded the house that bears their name, which still exists in Paris. Auguste Rivière was the director of the Luxembourg Gardens in Paris and the Hamma Experimental Garden in Algiers.

catch the epiphytes that flourished on the high branches of trees, or to cut inaccessible twigs. An articulated metal rod eliminated the need to carry this long pole or make a new one for each outing. A pellet attached to a rope, itself attached to a saw, could be thrown. The saw was stabilized by the second rope attached to it on the other side. The curvature of the saw was calculated to remain perpendicular to the branch when the two ropes were manipulated to saw it.

The more agile botanists did not hesitate to climb trees (the more cautious sent natives in their place), while some pulled plants with slingshots, or even rifles, to cut twigs at height. Tropical botany is not without danger, and here the gun is not to blame. It is important to be aware of thorny, irritating plants or those with caustic latex; a thick glove should be taken, which may protect from the bite of a tree snake.

In the wet season, insects and fungi devour herbaria. To prevent this disaster, it is sometimes necessary to activate the drying process by wrapping the plants over a fire, then keeping the herbarium dry and aired. Botanists also know a natural method: placing plants under a mattress overnight and airing them out every morning. To limit attacks by pests once the herbarium has been made, the plates are poisoned with various toxic products (mercury bichloride is often used). The operation is repeated regularly. The same applies to herbaria stored in temperate regions (where frost must also be avoided). Label inks must be resistant to time and humidity. China and gallnut inks are recommended. Gall is a growth produced by the stinging of a leaf blade by an insect. It is rich in tannins, and the substance derived from it is half-animal, half-plant, since the gall contains one or more larvae. The gallnut has been known to tanners, dyers and apothecaries in the Mediterranean countries since Antiquity.

In the tropics, it is advisable to store herbarium packets in metal boxes and visit them regularly. Fleshy fruits, flowers or succulents can be put in jars, but the results are not guaranteed, especially if the technique has not been mastered, as we have seen. In any case, it is always a good idea to take along a harvest notebook and a good pencil stroke. Such was the case with Eugenius; his drawings are remarkable for their finesse and precision.

Transporting live and dead specimens, some in jars, to the port by horse and mule over steep paths was another ordeal he had to overcome. After that, everything had to be shipped and stored away from sea spray and heavy seas. For live plants, highly prized in European botanical gardens, the simplest and least costly – but least secure – involved storing them in "gardener's crates". These are uncovered boxes with soil. Humboldt, whose personal fortune had enabled him to charter a ship and crew to travel to equinoctial America, was able to give instructions on storage conditions and had the right equipment at his disposal. For Eugenius, a mere passenger, things were undoubtedly more difficult. Could a young man of 25, with no title or status, have imposed strict compliance with any instructions?

Despite the precautions taken, in the 19th century there were enormous losses due to seawater reaching the plants and impregnating the soil. A competition with a prize was even organized by the English West India company. The winners were those who brought back live mangosteen and breadfruit plants. The mangosteen fruit is juicy and fragrant. It is highly prized, as is the round or oblong, cream-colored breadfruit. This competition stimulated research, of which the closed or screened boxes were one of the most remarkable results. Sides and tops were made of archal wire (paper or cotton-covered metal wire), with flaps that slid up and down at will to protect plants from bad weather. Wire panels allowed plants to be watered.

From the early 1830s onwards, Ward-style crates, portable or travel greenhouses, were used, in which plants could survive five months at sea. They resembled small wooden houses, consisting of a rectangular parallelepiped topped by an inverted V-shaped roof (112 cm × 48 cm, 80 cm high). One side of the roof was glazed and protected by wire mesh to prevent breakage. Dr. Ward was an English doctor with a passion for botany, and the inventor of these forerunners to modern terrariums. He wanted to protect his precious plantings from London's air pollution, caused by coal fumes and acid rain. Initially intended for the herbaria of middle-class homes, Ward's crates revolutionized long-distance plant transport. Thanks to this innovation, the botanical garden at Kew, London, successfully shipped six times more plants to the British colonies between 1832 and 1847 than in the previous hundred years.

This engineering, these techniques, these tricks and various skills, added to the professionalization of naturalist practices driven by museum instructions, were at the service of science, commerce and colonization.

Eugenius embarked on this adventure alone, albeit with the support and confidence of two compatriots and renowned professors: Lund and Reinhardt.

4.5. Flora of Central Brazil

The quality of his flora of Central Brazil, published from 1867 – he was only 26 years old – was acclaimed by specialists. The Danish journal *Videnskabelige meddellelser* (*Scientific Communication*) published it in several parts, as did the bulletin of the Natural History Society of Copenhagen in which Reinhardt wrote frequently (Warming 1877–1878, 1882, 1883, 1889, 1890, 1891, 1893). *The Société botanique de France* followed his advances, which were mentioned at its meetings.

Eugenius was a hard worker and a young man in a hurry, who also knew how to thank his predecessors and collaborators. In the introduction to Volume 1, he paid tribute to Lund, whose botanical explorations he cited from 1825 to 1827, then with Riedel in 1833–1835. These, he wrote, paved the way for his own field studies from 1863 to 1866.

Eugenius enlisted the help of Hampe, a Blankenburg pharmacist and moss specialist, to write the first book to emerge from his work at Lagoa Santa, with the aim of discovering the flora of Central Brazil. In return, he gave Hampe the benefit of his collection of Minas Gerais mosses, which enabled the German botanist to discover new species. Glaziou, director of the Botanical Garden of Rio de Janeiro, with whom he maintained an ongoing correspondence, sent him some Brazilian species (Hampe 1879).

It could be assumed that Eugenius had consulted the first volumes and fascicules of *Flora Brasiliensis*, which were in the process of being published, either in the Lund library or that he could have obtained during his stay in Rio. However, in the introduction to *Lagoa Santa,* dated January 1, 1891, he wrote that as a young botanist just starting out, he knew nothing about tropical plants and had only benefited from Lund's experience and collections – which was not bad – and from two general works: *Genera Plantarum secundum ordines naturales disposita* (1836–1850) by the Austrian Endlicher. *The Vegetable Kingdom*; *or the Structure, Classifications and Uses of Plants* (1st edition in 1830, 3rd in 1863) by British author Lindsey (Warming 1892b, p. 10). He made no mention of *Nova Genera et Species Plantarum Brasiliensium* (1823–1832, 3 volumes) or *Icones selectae Plantarum Cryptogamicarum Brasiliensium* (1827), published by Munich botanist von Martius.

While he continued to work on his own flora, a few years after his return to Denmark, publication of the *Flora Brasiliensis* volumes came to a halt. Eugenius supported a petition launched at the *International Congress of Botany and Horticulture* held in Paris in August 1878, at the same time as the Universal Exhibition. It was "addressed to H.M. [His Majesty] the Emperor of Brazil [Dom Pedro] to get the Brazilian government to continue the publication of *Flora Brasiliensis*". From Copenhagen, keeping in touch with the director of the Imperial Gardens in Rio de Janeiro, he was able to announce to the *Société botanique de France* in a letter dated May 4, 1879: "The Brazilian government will continue the publication".

Beginning in 1840 and reaching completion in 1906, sponsored by the Emperors of Austria and Brazil, and by the King of Bavaria, there were many contributors (over 60) including Lund, Claussen and Eugenius (around 20 citations). As early as the late 1860s, he sent one of the two co-authors, Eichler, his drawings and manuscripts on plants of the Cactaceae family (cacti) collected in Minas Gerais[10].

Completed after more than 60 years of effort, the monumental flora contained 15 volumes, 40 parts, 130 fascicles totaling more than 10,000 pages (more than 22,700 species). Its publication was hailed by the major scientific institutions in Europe. To this day, it remains the only complete flora of Brazil (Poeppig and Endlicher 1835; Martius and von Eichler 1840–1906; Glaziou 1905–1906; Davy de Virville 1954). It contains some very fine engravings that give a fairly accurate idea of the physiognomy of the landscapes depicted and of human activities (coffee and agave cultivation, mining, etc.). These included mule tracks comparable to those Eugenius must have taken in Minas Gerais.

What advantage could his own local flora have in a collective undertaking as ambitious as *Flora Brasiliensis*? What did Eugenius have to gain by embarking on such a project?

After all, the young botanist could have contented himself with contributing to the composition of the great Brazilian flora – which he did – thus appearing alongside those who were involved in tropical botany. It was a good springboard into his career. In fact, there was no redundancy or competition between his Central Brazilian flora and *Flora Brasiliensis*, nor was there any strategic error – quite the contrary. By the time he resumed his studies in Denmark and then Germany, his reputation as a botanist specializing in tropical flora had already preceded him.

4.6. An ecological study model

4.6.1. *Pioneering work*

After his flora of Central Brazil, Eugenius' second work was the *Lagoa Santa. And Bidrag til den biologiske Plantegeografi* (A contribution to the biological geography of plants), published in Copenhagen 25 years after his return from Brazil.

It soon became clear that his project went far beyond a simple inventory of a territory's flora. In 1908, the Portuguese translator wrote in his foreword:

> Dr. Eugen Warming's book represents the first attempt to organize a local flora for a specific region of a large Brazilian territory. The merit

10 Letters from Eichler to Engelmann, December 30, 1869 and February 16, 1875. His contributions also cover other families: Vochysiaceae, Annonaceae, Rosaceae, Melastomataceae, Rubiaceae and Compositae.

> of this work, however, lies not only in its systematic collection, with an enumeration of known and new species, but also in its simple phytographic descriptions and geographical distributions. It is, above all, the first attempt at biological and physiological studies ever carried out in Brazil on the relationships of the plant mantle with the climate, with the soil and with man himself (Ferri 1973, Foreword).

It was not just a question of describing and determining the distribution areas of species (chorology), but rather a causal approach to botanical geography, focusing on a number of factors. Thus, he dealt with issues relating to geomorphology, soils and climates, then describing plant formations before detailing their floristic composition and the impact of human activities.

He distinguished four main formations: forests, *campos*, sunlight-seeking formations and aquatic plant formations.

– Forests occupy depressions and valleys, following watercourses (gallery forests or riparian forests). They are characterized by the presence of lianas, sarmentaceous plants and shrubs that form the undergrowth. The flatter the terrain, the deeper the clay and the taller and straighter the trees, as they benefit from richer, more moist soil.

– *Campos* are open, partly grassy lands (grasses and tall herbs) with small shrubs and trees similar to mimosas, or flamboyants with spectacular red blossoms, in varying proportions[11]. He distinguished between *campos cerrados* (closed, thick) and *campos limpos* (clean, or green), thus introducing subdivisions that Lund had not identified. *Campos* are types of savannah. The word is derived from Spanish (*çavana*, *zabana* or *sabana*), defined in 1535 by the Spanish historian Gonzalo Fernández de Oviedo. It originally applied to the Venezuelan *llanos* and Brazilian *cerrados.* Oviedo defined this formation as follows: “The name savannah is given to land that is treeless, but with much high or low grass” (Oviedo and Valdès 1851–1855, p. 183). In South America, they are considered local types of savannah, with that of tropical Africa becoming the benchmark.

When Eugenius discovered these formations, they had been profoundly altered by human activity. Herders grazed their flocks, while farmers practiced slash-and-burn cultivation.

In 1864, he photographed a *cerrado* shrub (*Piptadenia macrocarpa)* that grew to over 10 m tall and resembled the mimosa. This species partly determines the physiognomy of the formation. The *campos limpos* are poorer in terms of flora. The

---

11 *Andropogon villosus, Rhynchospora warmingii* (a Cyperaceae dedicated to him), Cyperaceae, Compositae, Papilionaceae, Caesalpiniaceae, Mimosaceae, etc.

best *cerrados* are found on deep clay, but can also be found on sandy soils. If the floristic differences did not lead to physiognomic change, Eugenius considered that the name of the formation should remain unchanged. In other words, he referred more to the vegetation than to the flora, even though he had become a specialist in the latter for the area under study.

– Heliophilous (sunlight-seeking) formations are those found in the marshes along lakes and streams. Photographs taken by Eugenius clearly show the marshy banks of the Lagoa Santa.

– Finally, he distinguished the formations of limnophilous aquatic plants, which prosper in calm streams or stagnant water.

These last two formations play a lesser role in terms of physiognomy compared with forests and *campos* whose trees, shrubs and tall grasses mark the landscape.

The book discusses each formation in detail, treating them in relation to each other (Warming 1892b, Chapter 11, p. 350 *ff*) and to the seasons (*ibid.*, Chapter 12, p. 385 *ff*). Eugenius also mentioned vertebrates (Chapter 14), warning that he was not the author of these data. Finally, he reviewed the literature devoted to Lagoa Santa and its environment (*ibid.*, Chapter 15, p. 448 *ff*).

His work was therefore a work of ecology, dealing with both the plant and animal worlds, and their interactions. Interactions refer to the relationships between living beings themselves and their environment.

4.6.2. *A long history*

His approach to botanical geography is part of a history which, from the beginning of the 19th century, gave rise to two traditions, physiognomic and floristic, at the origin of which was the great traveling naturalist Alexander von Humboldt. Returning from his five-year voyage to equinoctial America with Aimé Bonpland, a botanist from La Rochelle (1799–1804), he published *Essai sur la géographie des plantes*, the foundation of the discipline known as botanical geography (or phytogeography), outlining a program that would keep researchers busy for the rest of the century.

Humboldt wrote on the physiognomic approach to botanic geography:

> In the variety of plants that cover our planet, we can easily distinguish a few general forms to which most of the others are reduced, and which present as many families or groups more or less analogous to each other. I shall limit myself to naming fifteen of these groups,

> whose physiognomy provides an important study for the landscape painter: 1. the shape of the scitamines (*musa, heliconia, strelitzia*) [tall herbaceous plants of warm regions, including banana trees, birds of paradise]; 2. that of the palm trees; 3. the tree ferns; 4. the shape of the *arum, pothos* and *dracontium*; 5. that of the firs (*taxus, pinus*); 6. all the *folia acerosa*; 7. that of tamarinds (*mimosa, gleditsia, porliera*); 8. that of malvaceae (*sterculia, hibiscus, ochroma, cavanillesia*); 9. that of lianas (*vitis, paullinia*); 10. that of orchids (*epidendrum, serapias*); 11. barbary figs (*cactus*); 12. casuarinas (ironwood), *equisetum* (horsetail); 13. grasses; 14. mosses; 15. lichens.
>
> These physiognomic divisions have almost nothing in common with those that botanists have made to date according to very different principles. We are concerned here only with the broad outlines that determine the physiognomy of vegetation. Descriptive botany, on the other hand, groups plants according to the affinity of the smallest but most essential parts of the fruiting body. It would be an undertaking worthy of a distinguished artist to study, not in greenhouses and botanical books, but in nature itself, the physiognomy of the groups of plants I have listed (Humboldt 1807, pp. 31–32).

Eugenius was clearly part of the physiognomic tradition. At first glance, it leaves room for a form of experiential knowledge, even a certain intuition built up by confrontation with different landscapes, without the need to be a high-level botanist. The geographer's eye is welcome, and even that of the artist.

Today, we need only mention the savannah to visualize a form of vegetation and landscape popularized by a large number of wildlife films about tropical Africa. If we go into a little more detail or, better still, if we walk the terrain, we can sketch out a typology based on basic botanical knowledge, leading to a distinction between the grassy savannah dominated by grasses between 2 and 6 m, wooded savannah (baobabs, acacias), park savannah (trees arranged in clumps), shrub savannah (dotted with shrubs less than 8 m high) and bushy savannah.

During the long mule-riding journey to Lagoa Santa, Eugenius built up mental images of the landscapes he crossed, recording his feelings. He also began to identify the region's physiognomic features, thanks to the naturalist's eye and methodology he had already acquired from studying the plant landscapes of his native Jutland. However, Eugenius' botanical geography did not neglect flora. In this, he was also linked to the floristic tradition, the foundations of which were also laid by Humboldt in the same *Essai sur la géographie des plantes*:

> Although the phenomenon of social plants seems to belong primarily to temperate zones, the tropics offer several examples. On the back of the long Andes chain, at a height of 3,000 meters, we find *brathis juniperina*, *jarava* (a genus of grasses close to *papporophorum*), *escallonia myrtilloides* [an evergreen shrub], several species of *molinia* [a herbaceous plant with purple spikes], and above all *tourrettia*, whose pith provides a food that indigent Indians sometimes dispute with bears. In the plains between the Amazon River and the Chin chipe, *croton argenteum* [cultivated in the King's garden], *bougainvillea* and *godoya* [hardwood shrub with yellow flowers] are found together; as in the Orinoco savannahs, *mauritia* palm, sensitive herbaceous plants and *kyllingia* [tropical herb]. In the kingdom of New Grenada, *bambusa* [bamboo] and *heliconia* [brightly colored flowers, food for hummingbirds] offer uniform bands uninterrupted by other plants: but these associations of plants of the same species are less extensive, less numerous, than in temperate climates (Humboldt 1807, pp. 18–19).

We are talking here about "social plants", "associations" in which plants are "gathered in society like ants and bees". The botanist comes to the fore, for example, when characterizing the "association of *Erica vulgaris*, *Erica tetralix*, *Lichen icmadophila* and *hoematomma*" (heathers and lichens) (Humboldt 1807, p. 15, p. 17).

The Swiss Augustin Pyramus de Candolle developed a research program for botanists in line with this floristic tradition. The aim was to identify plant communities on the basis of comparisons between inventories that were as exhaustive as possible, in order to identify characteristic species associations, statistically the most present in a given environment, even if they have a lesser impact on landscape physiognomy (de Candolle 1820, p. 384). For example, *Scilla lilio-hyacinthus* (lily-jacinth scilla) is a characteristic species of association of the submontane beech fir forest of Auvergne, represented around Lac Pavin. While it is clear that beech and fir dominate the overall landscape, this is not the case for this herbaceous plant, which, although it can reach a height of 1 m, only flowers in the undergrowth in April–May. But its presence alone characterizes the association.

Many authors, as for the physiognomic tradition, enriched the floristic tradition throughout the century, until the synthesis created by Eugenius at the end of the 19th century.

4.6.3. *Observe, measure*

As a field naturalist, Eugenius returned repeatedly to the same locations at different times of the year. He was able to note that the start of the new leaf germination phase, for most species, occurred at the transition between the dry and rainy seasons. The loss of leaves seemed to him to represent the best strategy for saving water. He noted that the pilosity, the thick felting, the glossy leaves, leathery, vertically oriented, reduced, the glandulous, star-shaped hairs, etc. all testified to the plants' response to the dry climate.

In particular, he cited *Anona furfuracea* (synonym *Duguetia furfuracea*), a slow-growing shrub 1–2 m high, highly resistant to drought and full sun, whose edible fruit has the consistency and taste of carrot cake. A branch collected in Lagoa Santa by Eugenius is kept in the herbarium of the *Muséum national d'histoire naturelle* in Paris. Another was brought back from "the *campo* near Pahna in the locality of St-Paul, Brazil" by Auguste de Saint-Hilaire, first published in his *Flora Brasiliae meridionalis*, 1824–1833, and preserved at the University of Montpellier.

Eugenius was from northern Europe. He appreciated the "sometimes exquisite freshness" of the forests. There, the species have larger leaves than those of the same family found on the *campos* where drought can cause leaves to fall, preventing transpiration. Nature brought him unfamiliar sounds. During the dry season, for example, some trees make a jingling or rattling sound as their leaves rattle in the breeze. When the rains come, brought by the prevailing west and northwesterly winds, "the many voices of amphibians" are heard.

After his first dry season, he was less enthusiastic than when he first arrived in Lagoa Santa:

> The area, like all the countryside, is not pretty. The view does not make a pleasant impression. The tone of the landscape is almost always grayish and sad, and, if spring has not enlivened everything with its cheerful greenery and flooded the fields with thousands of flowers, an imprint of desolation and death seems imprinted on the whole region (Warming 1892b, p. 16).

However, he referred to Lagoa Santa as the place "where I spent such happy, carefree years" (Warming 1892b, p. 10).

His repeated observations helped him understand vegetation cycles. In forests, litter production peaks from June to September, when there is a water deficit. The leaves forming the litter are responsible for returning nutrients to the soil. Flowering and fruiting are less seasonal, but also increase with the transition from dry to rainy seasons. He recorded all of these phenological changes (phases of development at different times), collecting, photographing, drawing and identifying, amassing a significant quantity of samples and data that he would exploit for decades. At the end of the century, these data were used in his treatise on ecology about the ways in which plants adapt to environmental conditions and the identification of the world's plant communities.

There are a number of elements that make it possible to identify the foundations on which his physiognomic botanical geography was based.

He drew on the meteorological data collected by Lund before him, and added his own measurements and observations. He recorded temperatures twice a day, in the morning between 6.00 and 6.30 a.m., and in the afternoon between 2.00 and 2.30 p.m. Across around a dozen pages, in *Lagoa Santa*, he provided data with monthly averages, maximum and minimum temperatures. The highest temperatures were measured between November and January (around 35°C for maximums and 15°C for minimums), and the lowest in August–September (around 30°C for maximums and 10°C for minimums). January averaged 24°C, July 16°C.

As soon as he arrived – in the dry season – he noted that plants were seeking shade between the limestone boulders that had collapsed at the foot of the vertical walls. Forest vegetation grows between these rocks. The ravines formed by the rains are often wetter. Over time, mule tracks may have formed here. Using them himself on a regular basis, he noted that the more they were used, the deeper they were. After several years, the rains widened those that remained damp, clammy and clay-scented. But most were narrow – around 30 cm wide – as the mules followed each other in line (Warming 1892b, p. 20, pp. 23–26).

He then identified and prioritized the factors affecting vegetation:

> The physical conditions that determine the clear separation of forests and *campos* are due to differences in soil moisture: while the forest supports the watercourses, the *campos* occupy the higher, drier parts of the territory. I'm inclined to believe that the forest is also fond of limestone rocks, because it finds there, and particularly at the foot of the rocks, a more considerable humidity; whereas, at the top of the rocks, this humidity is rather much lower. The hygrometric state of the air is also higher above the forest than in the campos (Warming 1892b, p. 318).

And 10 pages later:

> Water is a key factor in the development of the biological character of vegetation formations. Hygrometric conditions essentially determine the anatomical as well as the morphological structure of plants: the flora of Lagoa Santa is just one example. It's certain that the various tree forms in *campos* and forests are partly related to the degree of humidity of the soil and atmosphere. The same applies to the thickness of the bark, as mentioned above. The presence of numerous lignified, subterranean axes, which are found in *campos* vegetation and not in forest or other vegetation, must be at least partially attributed to physical differences (see chapter on fires in the *campos*). The extreme scarcity of creeping rhizomes or stolons in *campos* plants is due to the hardness of the soil; on the other hand, these plants are already more numerous in the loose soil of the forest, and even more numerous in marshy or generally damp soil. The *Campos* have almost no species with creeping or radicating aerial shoots; forests, on the other hand, have many, most of which grow on moist soil, because loose, moist soil obviously favors the development of adventitious roots (Warming 1892b, p. 329).

This attention to the aerial and subterranean organs that ensure the plant's survival, and this attempt to classify plants according to the constraints of the environment were taken up by one of his pupils, Christen Christiansen Raunkiær, who succeeded him as professor at the University of Copenhagen. In 1904, he developed a classification of biological types, which was subsequently refined. This classification is applied to the parts of the plant that survive the unfavorable winter or dry season, depending on the climatic zone. This is the case for "underground lignified axes rains", rhizomes and roots (Warming 1884, 1895, Chapter 2, 1923; Raunkiær 1904, 1905).

Finally, for Eugenius, water was the primary factor determining the development of plant forms. He marked the anatomical structure (internal) and morphological characteristics (external) of plants. The flora of Lagoa Santa provides daily proof of this. Secondly, the nature of the soil intervenes and, in some cases, humans have modified the natural evolution of vegetation.

4.6.4. *Reflecting on farming practices*

Eugenius was interested in the cultivation of *Coffea arabica*. He noted its flowering from September to November, at intervals of one or more weeks. He marveled: "On the same day, the coffee plants of the region are covered with fragrant white flowers; but the flowering lasts only two or three days, new flowers blooming after a certain interval" (Warming 1892b, pp. 404–405).

Certain farming practices attracted his attention, insofar as they heavily impacted the natural plant formations. In *Lagoa Santa,* for example, an 1864 sketch depicts a mountainous landscape with forested valley bottoms of the Rio das Velhas, the main tributary of Brazil's longest river (*Rio São Francisco*). Its course runs south–north through Belo Horizonte. In the distance, two plumes of smoke indicate forest fires. In this case, the aim is to gain arable land. Once the land has been conquered, fires are commonplace from July to September. It has destructive effects on flora and fauna. Human activities are therefore integrated into his study of botanical geography. He discusses the distinction between natural formations, modified to an extent that is difficult to define, and highly modified (secondary) formations.

Figure 4.5. *Landscape around Lagoa Santa towards the Rio das Vehlas (its river valley lies behind the two large hills). In the distance, the ruins of campos and forest fires can be seen (Warming (1892b, p. 16) reprinted in Warming and Graebner (1918, p. 344))*

Although he deplored the damage caused by "controlled burning" (clearing by fire) in the Brazilian *campos*, he understood its benefits for farmers, as it encourages the regrowth of cultivated species. This practice predates the arrival of Europeans. Eugenius recalled that this point had already been made by Humboldt. Species that survive this treatment have adapted. In the *campos*, the thick, woody and irregular underground organs are related to the dry climate, stoloniferous plants are lacking due to the hardness of the soil, but certain characteristics in the structure of the plants and in the physiognomy of the plant formations are fire-related. Thus, dry *campos* have vegetation whose stunted appearance is due to the fires. He noted the presence of a majority of perennial species adapted to drought, soil hardness and fire.

He then hypothesized that fire could be responsible for changes in the composition and distribution of plant formations. This led him to ask what he himself described as a fundamental question:

> Did the fires in the *campos* determine the formation of new species? Although this is one of the most important questions, I am forced to leave it unanswered […]. I believe that the frequency of fires may determine the transformation of a dry forest into a *cerrado*, and that land cleared in this way may, in a pinch, present a soil favorable to the immigration of *campos* trees (Warming 1892b, p. 274).

While Lund was imprudent in the eyes of his contemporaries when he put forward what was considered a bold hypothesis, Eugenius showed great reserve – to say the least – while still allowing himself to raise the question of the formation of new species. But did he really have a choice after Lamarck's publications (*Philosophie zoologique*, 1809) and those of Darwin (*On the Origin of Species*, 1859) on the transformation of species over time, and their evolution through natural selection? Here, Eugenius attributed the appearance of new species not to their transformation, but to the immigration of species from the *campos*. In other words, he bypassed the question he had himself formulated.

Conclusion to Part 1

A youthful journey interrupted the studies of a student spotted by Johannes Theodor Reinhard, professor of zoology at the University of Copenhagen. He proposed that he become, as he himself had been, secretary to a fellow Dane conducting paleontological research in Brazil.

A pastor's son with an excellent education and a passion for botany was no doubt already thinking about his future career. Reinhardt was a valuable supporter, a scientist renowned for his work on insects and terrestrial vertebrates. He had hinted that a botany project, initiated by Lund, was waiting.

Eugenius agreed to leave everything behind, despite the risks and the long absence involved, putting his university education on hold. He joined the ranks of naturalists making their "voyage around the world". Between Reinhardt's proposal and his departure, less than three months passed. Three months to make his decision, prepare for his trip and his stay, and say goodbye to his mother. Eugenius' diary shows that she was always at her only son's side. We may think that she let him go with emotion, but encouraged him.

By the time of his return, Eugenius had already established himself as a reliable botanist. He himself noted the positive reception of his first works. Twenty-five years later, he noted that his studies on the flora of Central Brazil were mentioned by more than 50 botanists.

Lagoa Santa became a model for Professor Warming's students, including Raunkiær, Frederik Christian Emil Børgesen and Ove Vilhelm Paulsen. The latter two published a study on the vegetation of the Danish West Indies, drawing on contributions from an expedition to Venezuela and the West Indies organized by

Eugenius in 1891–1892 (Warming 1893a, p. 21; Børgesen and Paulsen 1898, 1900)[1].

They followed the structure of the master's work: identification of ecological factors; presentation and classification of the main types of vegetation; description of plant morphology (leaves, stems, roots, etc.) in relation to their habitat; comparisons between formations in tropical and temperate regions. Photographs gave an idea of plant physiognomy and landscapes; line drawings provided details highlighting biological, morphological and anatomical adaptations to environmental conditions.

Eugenius' publications are methodological, conceptual and formal models.

For Eugenius, his stay in Brazil was like a life-size experiment, a means of forging a methodology and drawing comparisons with what he already knew, Danish flora and vegetation, and later with others.

His analogical thinking led him to "attach to the same physiognomic group [...] the *catinga* forests [a type of low forest made up of spiny, undeveloped trees in northeastern Brazil] and the *catanduva* [several species of trees in the genus *Pitadenia*]". Coastal forests and alpine *campos* are also related.

A year after the publication of *Lagoa Santa*, he wrote:

> If we examine other regions of South America, such as the *pampas* and *llanos* of Venezuela and the savannahs of Guyana and Venezuela, we can see that the soil of the Venezuelan *pampas* and *llanos* is developed on geologically much younger ground than that of the *campos* of Minas Gerais. As a consequence of this difference in soil age, the savannahs of Caracas and Valencia (Venezuela) have a flora similar to that of the *campos* of Lagoa Santa, but much poorer. As for the savannahs of French Guiana, judging by Schomburgk's descriptions, they correspond perfectly to the *campos* of Brazil; the soil is also very old and the flora is probably no less rich (Warming 1893a, pp. 156–157).

These physiognomic and floristic observations are complemented by reflections relating to phenological differences between plants growing in different climatic conditions:

1 Børgesen wrote his thesis on marine algae in the Faroe Islands (1904).

> If, from Denmark, my homeland, we move northwards, the development of vegetation follows an increasingly rapid march; the main time arrives later, but foliage, flowering and fruiting must, because of the shorter duration of the vegetation period, be completed more quickly. The opposite phenomenon occurs as we move southwards. We have almost no data relating to what happens in the tropics, despite the ease with which many botanists can make such observations. From the few notes I have been able to take on this subject, it seems certain to me that alongside certain plants which develop their fruits as quickly as ours, many others, and especially the woody plants, extend the ripening period of their fruits beyond several months and up to a year (Warming 1892b, p. 314, p. 335).

According to tropical ecologist Robert J. Goodland, *Lagoa Santa* represents "a paradigm of ecological study", and Eugenius was the founder of tropical ecology. (Goodland 1975, p. 244). Brazilian biologist Felipe Costa considered that *Lagoa Santa* founded tropical plant ecology. It is a "pioneering work in the history of community ecology" (Costa 2005). For Löfgren, a Swedish botanist who lived in Brazil, Eugenius published the first biological and physiological study of Brazil on the relationship between vegetation cover and climate (Ferri 1973).

It has sometimes been written that Eugenius had a static approach to ecology, characteristic of that developed in Europe during the 19th century. For so much, he also delivered observations on vegetation dynamics. For example, he described the succession of vegetation on a corn crop abandoned after two years. He noted the appearance of weeds, then shrubs and finally trees. "In twenty or thirty years, the forest was reconstituted, not in its primitive form, but in a *capoeira* state"[2] (Warming 1892b, p. 319). *Capoeira* is a strip of Amazonian-type forest along the coast.

The small Brazilian territory, so thoroughly explored, enabled Warming to draw comparisons with other tropical regions and to initiate generalizations made possible by his physiognomic and causal approach to botanical geography, the foundation of tropical ecology.

Ecology, like natural history in general, was practiced in tropical regions by travelers passing through or staying for varying lengths of time, who exploited and published their research in Europe and non-tropical America. They exported their harvests and created collections to scientific centers in the North, even though the first research stations in tropical regions were created in the 19th century. We could

2 *Solanum*, *Lantana*, *Croton*, *Sida*, etc.

mention the garden laboratory at Buitenzorg (Java), an important scientific center (Lachmann 1884–1885; Raby 2017, p. 8, pp. 25–27)[3].

In the 1960s, the great ecology specialist Ramon Margalef considered the success of the ecological schools in Europe, North America and Russia, regretting "that the tropical rainforest, the most complete and complex model of an ecosystem is not a suitable place for ecologists to breed [specialists in ecology]" (Margalef 1968, p. 26).

A century after the publication of *Lagoa Santa*, an exhibition of photographs taken by Eugenius was organized at the Botanical Garden of the University of Copenhagen by plant physiology specialist Aldo Luiz Klein, from the Faculty of Science and Letters on the Assis campus (São Paulo State).

A conference also organized by Klein was dedicated to the man who is recognized in Brazil as one of the most important European scientists to have lived there. To mark the occasion, he collected previously unpublished items (herbarium plates, travel notebooks, etc.) and set up a traveling exhibition (São Paulo, Assis, Campinas, Santos, Rio Claro). A book published in 2000 contains elements of this commemorative exhibition and biographical data on Eugenius.

For current authors of works on the plant ecology of the Lagoa Santa region, reference to Eugenius' tropical work is still the rule, and his work features in the basic bibliographies of courses given at the State University of Minas Gerais.

As for Eugenius, after his first contact with the tropics, reignited by other later trips to these latitudes, he worked all his life on the data he had accumulated. In November 1923, five months before his death, he again evoked this period spent in Brazil in two lectures given for the Danish Botanical Society, of which he was an honorary member (Klein 2000, p. 11).

The manuscript of these last conferences can be read:

> From time to time, I think of my happy, sunny days in Lagoa Santa – a tear forms at the thought that I will never see her again, that I will never hear the voice of the seriema on the deserted campos, that I will never let my gaze wander to the serra da Piedade, at the summit

3 *Annales du jardin botanique de Buitenzorg* was published from 1876 to 1940. Pathé frères released a silent film in 1914, produced by Oriental films: *Le Jardin botanique de Buitenzorg, île de Java.*

shrouded in clouds in the morning hours, far to the east, which I will never row in a canoe on the small lake [...]. Often, the little Brazilian melodies ring out again, and I long to revisit these places. Then I feel the truth of the old adage that "no one walks under palm trees with impunity". At Lagoa Santa, there was light, happiness and peace! (Prytz 1984, p. 176)[4].

4 Allusion to Goethe's *Elective Affinities* (1872): "You cannot walk among palm trees with impunity, and your sentiments must surely alter in a land where elephants and tigers are at home".

PART 2

Good Times and Bad

5

Outstanding Doctoral Studies

5.1. Exceptional conditions

Eugenius seemed anxious to return home and resume his studies at university, but he was delayed. He wrote from Kolding to the director of the Botanical Garden on October 23, 1866, that he would not be in the capital before November, and asked him to take care of the plants he had sent[1].

Soon after, the student was able to resume his interrupted studies and live comfortably thanks to his inheritance. In 1867, he gave a lecture on his excursion to the Serra da Piedade, and wrote to Lund that it was greatly appreciated. He lived in Copenhagen, at 26 Dosseringen overlooking Peblingesoen, which inspired the Danish painter Paul Gustav Fisher to paint skaters.

Eugenius reunited with some of his teachers, including Japetus Steenstrup, Ørsted and of course Reinhardt. He obtained his master's degree in 1868, moved to Munich, then to Bonn, from where he returned in 1871 to defend his thesis. Eugenius' family always chose the best for him. As an adult, he continued in this vein.

In the 1860s, the German Confederation boasted 20 universities and around 12,000 students. His focus was on the University of Munich (1826), the second largest in terms of student numbers (approximately 1,400), after Berlin (approximately 1,800). During this period, student numbers in philosophy (including the humanities and sciences) increased due to the diversification of disciplines and the attraction of the exact sciences (Dreyfus-Brisac 1878; Krebs 1984). These figures are not comparable with those of Denmark, where the only university had between 1,000 and 1,500 students in the 1880s. The capital of the Kingdom of Bavaria underwent

1 Letters preserved at the Uppsala University Library. See: https://www.alvin-portal.org/.

great cultural and scientific development under King Ludwig II. But above all, for Eugenius, it was home to the great botanical collections of von Martius, author of a three-volume flora of Brazil (1823–1832) and initiator of the monumental *Flora Brasiliensis*, on which he worked until his death in 1868. Eugenius had been contributing to the project since the early 1860s.

His arrival in Munich was preceded by various plant shipments. Plates from his herbarium enriched the collection managed by Ludwig Adolph Timotheus Radlkofer, a Bavarian botanist and taxonomist who had been teaching in Munich since 1859, and who also managed the botanical garden. With him, Eugenius practiced plant anatomy. He met botany professor Eichler, to whom he had already sent plant drawings and manuscripts, and Engler, who had received shares in his herbarium. Eugenius' flora of Central Brazil was beginning to be published, and was very well received. Finally, his name was also linked to that of Lund, many of whose works have been translated into German. The student thus presented himself with a more than respectable calling card.

Among other renowned scientists, the great Swiss botanist Karl Wilhelm Nägeli, professor in Munich since 1857, supervised his research during the summer and autumn of 1868. Physiologist and botanist, he worked in particular on cell division – he identified the division of the nucleus as an essential step – and pollination (Magnin-Gonze 2004, p. 163 *ff*). He opposed Darwin's evolutionary theory.

Eugenius went to Bonn in 1871, a few months after the creation of the German Empire, following the French defeat. He spent a semester there. The university was renowned for the quality of its teaching in the natural sciences. He completed his thesis under the supervision of Johannes Ludwig Emil Robert von Hanstein, a specialist in plant anatomy and morphology and professor since 1865. He headed the university's botanical institute, housed in Poppelsdorf Castle. It was bequeathed to the university founded in 1818 by Frederick William III, in the context of the reunion of Bonn and Prussia's Rhenish provinces after the Vienna congress (November 1814 to June 1815), at the origin of the Germanic Confederation, which essentially returned to the boundaries of the Holy Roman Empire. Hanstein's apartments were located in the castle. The building housed botanical, mineralogical and zoological collections, lecture and reading rooms, and a museum.

In line with Humboldt principles, professors were free to choose the subjects of their lectures and students were responsible for their own education. These principles laid the foundations of "academic freedom", even though universities, like other institutions, could be subject to censorship, particularly in the political arena. In any case, full professors had absolute power to judge the merit of a thesis and choose their own research program.

Eugenius attended Hanstein's seminars, which accompanied the preparation of his thesis. Hanstein worked on histogenesis (the formation of tissues during embryonic development) from undifferentiated cells of the apical meristem of plants – a zone of cell division that enables the growth and differentiation of tissues – and on the fertilization process (in this case, pollination) in ferns. Under his guidance, Eugenius trained in microscopic methods. Hanstein was the director of the botanical garden and greenhouses adjacent to Poppelsdorf Castle. Professors of botany were often entrusted with the management of their university's gardens. This was the case for Eugenius in Copenhagen.

In Europe, botanical gardens were essential to research. In Germany, they were almost all attached to universities. They were study and experimentation grounds for researchers, students, naturalists, physicians, pharmacists, horticulturists and agronomists. In France, in provincial towns, they could be managed by academic societies. Open to the public, they were also recreational areas and structuring elements of urban planning. From the second half of the 19th century onwards, they played a part in the greening of cities, driven by the hygiene movement (Drouin 1993; Fischer 1999; Matagne 2022).

Collections and exchanges with colonies enabled the introduction and testing of the acclimatization and naturalization of exotic plants, for economic and research purposes. Botanical gardens in tropical regions were rarely attached to scientific establishments. The exception was Buitenzorg (now Bogor) in Java, in the Dutch East Indies. It gave rise to remarkable studies cited by Eugenius. It brought together many species that were difficult to access, being scattered over several islands or collections requiring long and perilous expeditions. It was associated with a museum, a library, a test garden and an agricultural school. It offered excellent working conditions for botanists, and was regularly visited by a Dutch liner. Danish botanist Hajlmar Jensen returned from Java with photographs published in an edition of the *Lehrbuch* (Warming and Graebner 1918). Jensen was employed on tobacco plantations in Java.

Eugenius made an excellent choice. In Munich, then in Bonn, he benefited from a particularly favorable environment: modern facilities, collections, libraries, botanical gardens and high-caliber teaching staff.

In the year he earned his PhD, Eugenius married Johanna Margrethe Jespersen (known as Hanne Warming) on November 10, 1871. He wrote to Lund that she "is not really beautiful, but she is a pleasant and virtuous girl, who plays, sings, speaks French, etc." (Prytz 1984, p. 40). Her father, Jacob Vilhelm Jespersen, was a diocesan physician and later *justitsraad* (legal adviser), and her mother, Anna Marie Kirstine Bloch, was the third of seven siblings.

Marie, their first daughter, was born on October 2, 1872, Jens on December 9, 1873. Their eight children were born between 1872 and 1884[2]. The Warmings had three employees, one of whom was especially dedicated to assisting the young mother. Fro (1875–1880), Povl (1877–1878) and Svend (1879–1882) died in infancy. Svend's twin Inge, died aged 14 (1879–1893). Deeply affected by the death of his first son in 1878, Eugenius wrote to a correspondent that he had fallen behind with his mail (February 24, 1878). On April 4, 1880, he announced Fro's death. In 1894, still traumatized by Inge's death a year earlier, Hanne and Eugenius feared for the life of 10-year-old Louise, who appeared to have been struck down by diphtheria. Laboratory bacteriological analyses allayed their fears. All she had was a common sore throat. She had been infected by bacteria in the water (Prytz 1984, p. 58, p. 93). In 1878, the family moved into a house built in the new Rosenvængets (Østerbro) housing estate, at number 19 on the main street. The street was named after Copenhagen's first residential neighborhood, close to the beach, which had been divided into approximately 50 plots from the 1850s onwards.

Figure 5.1. *Portrait of Eugenius Warming in 1879 (public domain via Wikimedia Commons)*

5.2. Small objects, big discussions

His thesis, defended in Copenhagen in 1871, focused on the ontogeny of euphorbias, i.e. their development from fertilization to adulthood.

---

2 Marie (1872–1947), Jens (1873–1939), Fro (1875–1880), Povl (1877–1878), Svend (1879–1882), Inge (1879–1893), Johannes (1882–1970), Louise (1884–1964).

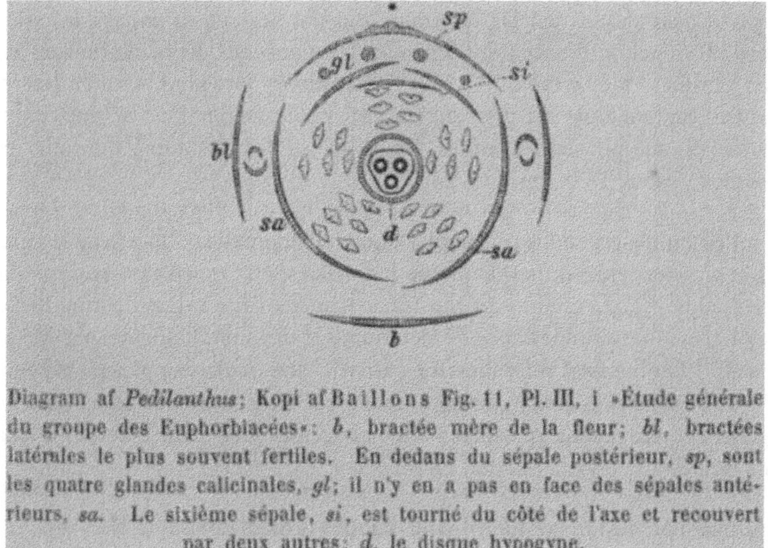

Figure 5.2. *Floral diagram of a euphorbia (*Pedilanthus*) according to Baillon (Warming 1871, p. 59)*

Euphorbias are latex plants, whose species are tropical and temperate. When Eugenius began to take an interest in them for his research, it was debated whether a cup-shaped organ (the cyath), characteristic of the Euphorbiaceae family, was a flower or an inflorescence, i.e. the grouping of very small flowers, without petals or sepals, on the same stem. Eugenius showed that this was an inflorescence of greatly reduced male flowers surrounding a female flower. This question may seem small, both literally and figuratively. However, it was linked to a field of research in which Eugenius was to excel: morphology and organogenetics (the formation and development of organs). His 1871 dissertation is subtitled "A morphological-organogenetic study" (Warming 1870b, 1871).

He explored the path opened up during his doctoral studies and began publishing in *Botanische Abhandlungen* (*Botanical Treatises*), a journal edited by Hanstein (Warming 1873; 1876). His work focused on the formation of pollen and anthers – the terminal part of the stamens – in Angiosperms (flowering and seed-bearing plants) and on the inflorescence of the Asteraceae (also called Compositae or Composaceae), whose tiny flowers are united to form a capitulum (as in the case of dandelions).

One of his contributions, dated 1878, concerns the ovules of seed plants (Warming 1878). It confirms the high level of expertise acquired since and thanks to his thesis. It also shows that Dr. Warming did not hesitate to engage in polemics. Published in French in the *Annales des sciences naturelles*, he made himself known to Julien Vesque – they were of the same generation – to whom he often referred in his treatise on ecology on the subject of plant adaptation to environmental conditions. Vesque praised Eugenius' "sagacity" in demonstrating the foliar nature of the ovule (Vesque 1878, pp. 239–240).

He relied on observations of samples showing anomalies: "In all cases carefully studied, it was recognized that the ovule had been transformed into a foliole, a leaf lobe". He adds: "The certainty of the foliar nature of the organ bearing the ovule [...] is of the utmost importance in determining the morphological nature of the ovule itself". The general rule, therefore, is that "the ovular nipple is [...] foliar in nature". This point was hotly debated among specialists, and referred back to the more general, ontogenetic question of the processes that lead embryonic cells to form the tissues and organs of adult reproducers.

Eugenius' conclusion conflicted with that of his main contradictor:

> Mr. Strasburger believes that he refutes this objection [the leaf origin of the ovule] by saying that he has demonstrated by homology and the history of development that the ovules of conifers are buds; that they must therefore be buds everywhere, even in plants where they are born on leaves [...]. One cannot conceal from oneself how suspect is this explanation based on a single exception (Warming 1878, pp. 197–198, p. 224).

Eugenius' contemporaries often remarked on one trait of his character. On first reading, it is clear that he was a divisive man. We will return to this when we discuss the period when he was a professor of botany in Copenhagen, after his return from Stockholm.

He systematically deconstructed Eduard Adolf Strasburger's demonstration. Strasburger was also a student in Bonn, receiving his doctorate in 1866 from the University of Jena where he was already a professor of botany. He took up this post in Bonn in 1880, becoming one of the pioneers of plant cytology (now cell biology).

Eugenius' research was based on numerous detailed observations, thanks to the production and study of thin sections. He benefited from modern equipment at Poppelsdorf Castle.

In the 19th century, the use of the microscope enabled great progress to be made in plant cell biology. The instrument was responsible for the recognition of cell division as a universal phenomenon in the 1840s. Many botanists embarked on microscopic studies, and the precision of their drawings became remarkable. High-performance microscopes were manufactured in Germany (Carl Zeiss in Jena) and in France (optics engineer Charles-Louis Chevalier, "Nachet et fils"), with lenses that no longer exhibited spherical aberrations. Previously, these had caused distortion, making it impossible to obtain a sharp image around the edges. A double-lens system (achromatic doublet) made it possible to compensate for these aberrations. Eugenius was also able to use immersion microscopes. The subject, immersed in a special oil with the same refractive index as water, increased image resolution for the high magnifications he required.

Another of Eugenius' hallmarks was his thorough, precise review of the works that paved the way for his research. Unlike other authors published in the *Annales des sciences naturelles*, he cannot be reproached for not taking into account this or that work deemed essential. His writing was very dense, with a structured plan, convincing demonstrations and a technical vocabulary, but didactic devices made the transitions visible and enabled the reader to follow the reasoning, to solicit his attention by regularly recalling the questions to which he intended to give answers. He indicates which questions remain unanswered, and which answers are incomplete or provisional.

He proceeds by accumulating observations and facts, giving the appearance of inductive reasoning. He criticized Strasburger for his preconceived ideas, based on his faith in a theory that he wanted to confirm, resulting in interpretations of phenomena that, in his opinion, were far removed from observed reality. Yet Eugenius himself fell into the trap he denounced: "I would be happy if this memoir could contribute to the early acceptance of Brongniart's theory as the only admissible and true one. If I am now convinced of this, I owe it largely to the ingenious Slavic botanist Ladislav Celakovsky" (Warming 1877, 1878, p. 255). Ladislav Josef Celakovsky suggested the leafy nature of the ovule through comparative studies. This was an essential basis for Eugenius' work.

French botanist and paleontologist Adolphe Brongniart mentioned by Eugenius – co-founder of the *Annales des sciences naturelles* – observed the formation of embryos from ovules in many species in the 1840s, after contact with the contents of the pollen tube, which carries "a spermatic fluid". In several species (e.g. *Primula sinensis*, the Chinese primrose), he remarked on the monstrous transformation of ovules into small, lobed leaves covered with short hairs, or stamens turning into pistils when close to the central axis (*Polemonium coeruleum*, or Greek valerian).

The theory Eugenius wanted to defend – the foliar theory of the flower – was presented by Brongniart, based on a few cases of plant monstrosities in the *Annales des sciences naturelles* (1844). Eugenius put into practice one of the author's propositions: "The examination of monstrosities can often enlighten us on the real nature of the various organs [of the flower], on their relations and on the analogy of the various parts which constitute them" (Brongniart 1844, p. 21). Thus, using the same materials and methods, Eugenius too was able to demonstrate the return of certain organs to their undifferentiated state, making it possible to trace their origin and identify homologies. Altered states represent different degrees of monstrosity, seen as deviations from normal development.

This research was related to plant teratology, which at the time was considered to be the study of monsters. It was based on Goethe's theory of metamorphosis. It was developed by several botanists, including Moquin-Tandon, de Saint-Hilaire, de Candolle and Brongniart. A question runs through this field of research: Are the changes observed "degenerations" or improvements, "glorifications"? The term metamorphosis used by Goethe is often preferred, as it remains neutral (Moquin-Tandon 1841, pp. 198–199).

According to Eugenius, the peculiarities of aquatic plants observed during his stay in Brazil "are to be interpreted in general as morphological and anatomical degenerations and retrogressions" compared to land plants. These retrogressions are adaptive (Warming 1896, pp. 95–96; 1902, pp. 99–100). As for monstrosities, he did not question their causes. That was not his subject. For him, it was a way of understanding the ontological origin of plant organs, in this case the ovum.

From ontogeny to phylogenesis it was just one step, taken by German Darwinian biologist Ernst Haeckel for the animal kingdom. Based on the study of vertebrate embryo development, he stated that ontogeny recapitulated phylogeny. Thus, during embryonic development, an organism would pass through stages representing those of ancestral species.

According to his colleagues, Eugenius paved the way for a new branch of botany, the "biological morphology" of plant organs. In an 1884 study, he developed a research program that he wanted to see adopted by Danish botanists. The aim was to link morphological studies with the study of relationships between species, their lifestyles, their migrations (Warming 1882–1886, pp. 139–140, 1884, pp. 59–64).

Eugenius temporarily abandoned organogenetic research to others. He was busy with other work. He returned to it after the publication of his *Oecology of Plants* (1909) (Warming 1914).

6

Intense Activity, a Detour via Stockholm

6.1. Intense activity

While in Brazil, Eugenius had already published a small study of "luminous phenomena in the plant world" (Warming 1864). On his return, articles that foreshadowed the future *Lagoa Santa* were published: "Representations of nature in the tropical regions of Brazil" (Warming 1867, 1868). He discussed the climatic conditions of Minas Gerais, which he compared with those of Denmark, describing the different physiognomies of the *campos,* their flora and fauna, and the agricultural practice of fires. He published his first botanical and landscape drawings of Brazil.

He was remarkably active. For a year, he was co-editor-in-chief of *Tidsskrift for Havevæsen*, a new fortnightly gardening journal, whose first issue appeared in April 1866. The periodical disseminated a wealth of information on plants, plant breeding, greenhouses, irrigation, botanical gardens as well as meteorological, economic and legislative information. It was edited by Jens Andreas Dybdahl, who, despite resistance from his colleagues, established himself as the leading horticulture professor at the agricultural college. Eugenius knew Dybdahl before he left for Brazil, as attested by two letters, the first from Rio de Janeiro dated May 13, 1863 – already mentioned as accompanying the shipment of botanical samples – the second from Lagoa Santa, dated April 25, 1865. These two letters also included the names of people with whom Eugenius would be in contact on his return to Copenhagen: Hjalmar Frederik Christian Kiærskou, editor-in-chief of *Botanisk Tidsskrift* (*Botanical Journal*), published by the Danish Botanical Society, in which Eugenius published regularly; Johan Martin Christian Lange, director of the botanical garden behind Charlottenborg Palace, with whom he corresponded from Brazil; August Weilbach, botanist and gardener of the same garden, and president of the Horticultural Society; Julius Jens Emil Hoffmeyer, journalist, botanist and librarian;

Professor Ørsted; and Didrik Ferdinand Didrichsen, known for his participation in a major Danish scientific expedition (1845–1847)[1].

In 1868, just before heading for Munich, Eugenius took part in a scientific conference organized by the Scandinavian Association of Naturalists, held in Christiania (Oslo). The only botanist in the group, he reported on his Brazilian collections. He was accompanied by Danish chemist Sophus Mads Jørgensen, whom he would meet again years later on the board of the Carlsberg Foundation.

The Scandinavian Conferences were a series of meetings (1839–1936) organized for Norwegian, Swedish and Danish scientists. Several hundred of them attended during the heyday of the Scandinavian movement. At the 1847 banquet in Copenhagen, Japetus Steenstrup was seen listening to the speaker, as depicted in a mural painted in the university classroom. Initially intended to report on research results, these meetings gradually lost importance with the boom in scientific publications, becoming mere moments for presenting summaries of studies already published. The meetings became less frequent. Eugenius remained assiduous. He attended the meetings in: 1873, 1880, 1892, 1898 and 1916.

Even before defending his doctoral thesis, he was building up his network of professional relationships. He strengthened this network over the decades, eventually becoming a network leader himself.

6.2. An initial positioning with regard to evolutionary theories

After Vaupell's death (1862) Eugenius replaced him as co-publisher of *Tidsskrift for populaere Fremstillinger af Naturvidenskaben*, already cited for the Darwinian views of its founder, zoologist Christian Frederik Lütken.

Lütken magazine was edited by Danish scientists and gave access to German, English and French authors by translating their writings. It was aimed at an educated readership. The theory of evolution was presented and discussed as early as the 1860s.

Eugenius was one of the first Danish naturalists to take a position on Darwin's theory. In a text published in 1870 on "The struggle for existence among plants", he drew on the work of the Dane Johannes Frederik Johnstrup on the geology and paleontology of Scandinavia, on those of the Darwinian Joseph Dalton Hooker from Edinburgh University for paleobotany (Warming 1870a; Hjermitslev 2011, p. 294).

1 Letters preserved by the Uppsala University Library. See: https://www.alvin-portal.org/.

While Johnstrup accepted the idea of evolution in the sense of Charles Lyell's principles of geology, he was not a Darwinian.

Eugenius had read the first English edition of *On the Origin of Species* (1859) (Warming 1870a, p. 370). He cited Linnaeus, Cuvier and Agassiz as representatives of an ancient theory that he considered outdated.

He presented the three pillars on which, in his opinion, Darwin's doctrine rested:

1) species are changeable;

2) changes could be transmitted from parents to children;

3) through the struggle for existence and natural selection, these qualities were fixed and new species formed through a series of generations.

But he believed that Darwin's followers – of which he was clearly not one – now had to prove the theory's validity.

For his part, drawing on the plant world, he outlined examples of advantageous characters that eventually took hold, but he pointed out: "To the extent that new qualities are inherited from generation to generation, they would become less and less variable, they become more and more fixed, and the formation of a new species over time from the first variety may be accidental". Thus, he was particularly interested in what he interpreted as a fixist dimension of Darwin's theory: new variations being fixed by heredity. The appearance of new species thus becoming accidental.

Through the struggle between species and natural selection, nature is able to produce ever more perfect forms, he concluded. In the human species, "the result can only be the same: the triumph of the intelligent and the gifted, of nature and man, and the constant development of humanity towards greater perfection, a more perfect bodily and mental organization". These reflections placed him outside Darwin's theory, as it implied that variations were random, not purposeful.

In botany, he acknowledged the concept of the struggle for life, hidden behind the seemingly peaceful life of plants. In fact, plant species struggle to conquer space. In the beech forest, for example, "the shade that refreshes us is the ruin of thousands of creatures". The variety of anemones, primroses, pulmonarias, larkspur, etc. disappears as soon as the leaves form and darken the undergrowth. Even young beech trees suffer, deprived of light by the leafy branches of their taller relatives. Spruce forests are even darker. Only the "creatures of darkness", i.e. "pale fungi that seek obscurity, cover the forest floor". "One growing forest succeeded another, in the course of [geological] time, one period of the Earth succeeded another, each with its particular vegetation".

However, he limited this phenomenon of the struggle for existence to situations where the ruin of one species favors another. In most cases, it is climatic factors that are to blame for the disappearance of species and their geographical distribution: "Climate is one of the principal factors of influence in the distribution of plants" (Warming 1870a, p. 351, p. 353, p. 356, p. 371, pp. 373–374). He also discussed the role of animals – insects in particular – the action of humans on the establishment or disappearance of species (cultivated fields or fields left fallow), the work of gardeners, acclimatization trials, etc.

He then laid the methodological foundations of his botanical geography, in terms of the theory of evolution:

– it is climate first and soil second that must be studied to understand the geographical distribution of plants;

– therefore, if the struggle for existence and competition play an important role, they are no longer primary. They manifested themselves in Scandinavia at the end of the Ice Age and with the emergence of the Danish islands.

Finally, the struggle for existence and competition have been driving forces throughout geological history. Today, the adaptation of plants thanks to special devices is one response to environmental conditions; migration, which he regarded as having been insufficiently studied, being another.

His project of ecological botanical geography was therefore consistent with an interpretation of phenomena that led him to downplay, without rejecting, the explicative value of Darwin's theory.

He made extensive reference to the botanist Nägeli, whom he met in Munich and who opposed Darwin's theory. It seems that he had a great influence on his thinking on evolution and, for ecology, drew his attention to the importance of studying plant communities. In fact, Eugenius urged botanists to be more interested in groupings than in competition, to pay less attention to rare species and more to common species, those found in communities. As early as the 1870s, the ecology of plant communities, central to his 1895 treatise, was already taking shape.

6.3. An unbearable situation

Before leaving for Brazil, Eugenius took no part in a controversy between professors Anders Sandoe Ørsted and Japetus Steenstrup. He was only a student at the time. Entomologist Jørgen Matthias Christian Schiødte, professor and inspector at the Copenhagen Museum, who led the fight against Steenstrup, mobilized a majority of young botanists and zoologists. By the early 1870s, it was clear that

Eugenius was on Steenstrup's side. Initially, this choice cost him a coveted university post.

When, on Ørsted's death on September 3, 1872, the position of professor of botany became vacant, Eugenius thought – indeed, he took it for granted – that it should be for him. But he was on the wrong side. His choice, driven by loyalty to his zoology professor, can be read as one of the rare strategic errors of his career. Support was sorely lacking, at a time when he needed it most.

A group of students and botanists led by Samsøe Lund, of whom Professor Didrichsen was a great supporter when he was an assistant at the botanical garden (Lund 1883)[2], caught the eye of Carl Christian Hall, a leading statesman during the difficult period of the Duchy War. Since 1870, he had been Minister of Public Worship (educational matters falling within his remit). Opponents of Eugenius urged the minister to press for the appointment of Ørsted's successor. He followed them, against the recommendations of the Faculty of Science, and Didrichsen became professor and director of the Botanical Garden, until his forced resignation in 1885. Indeed, it seems that he was asked to say farewell on September 1.

Much appreciated by his students, he knew how to make his lectures lively and stimulating. On the other hand, he ended up being unanimously opposed by his colleagues, and Ørsted became one of his strongest opponents in the 1860s. Didrichsen worked a lot but was a wanderer and published little. He left behind a large amount of disorganized notes. Eugenius judged his scientific output to be insignificant. His organizational plan for the new botanical garden, installed in 1871 and handed over to the university in 1874, on the site it still occupies today, was heavily criticized. Several other academics contributed to the plans, including Ørsted, Steenstrup and Lange, director of the previous botanical garden. The design committee for the new garden also included a number of prominent figures, including Jacob Christian Jacobsen, founder of the Carlsberg company and a major backer of the project.

Even though Eugenius was not a member of the committee, he took part in the discussion on the layout of the new garden and did not hesitate to formulate, vigorously, his propositions, in opposition to those of Didrichsen. The controversy escalated, and the major daily paper *Dagbladet* echoed it. The names Ørsted, Didrichsen, Dybdahl, Rasmussen (mentioned as a gardener) and Warming on the

---

2 He announced his plan to publish, under Didrichsen's authority, notes on the Copenhagen botanical garden and on various events involving it. From 1882 onwards, he was a *docent* in botany (equivalent to a lecturer) for future pharmacists, and archivist of the Copenhagen Botanical Society.

first line. The daily in which the players in the quarrel appeared was even quoted in *Botanisk Tidsskrift*, the journal published by the Botanical Society of Copenhagen (1880–1881, p. 170). The garden, inaugurated in 1874, boasts a complex of greenhouses – financed by the Carlsberg Foundation– built on the model of the Crystal Palace erected for the Great Exhibition of the Works of Industry of All Nations, in London in 1851.

A look at the correspondence between Eugenius and Didrichsen shows that relations between the two men deteriorated, whereas before his departure for Brazil, the student invited the professor, almost 30 years his senior, on botanical excursions (e.g. on June 6, 1862). The tone changed by the end of the 1860s: "I'll come and borrow books on Monday," Eugenius wrote to him curtly (July 18, 1869). A year later, he deplored the fact that they had not been able to meet on important issues (October 12, 1870). His requests, some of which were tantamount to seeking favors, were refused. On October 3, 1877, for example, Didrichsen responded that he would not allow anyone who did not hold a position in the botanical garden to use it at night by candlelight. His request to visit the library outside office hours went unheeded. He had to insist on ordering books for the library. Even if, logically, all this has to go through the principal, he found it hard to accept Didrichsen's whim to go to the botanical garden with his students, to get non-academic botanists to consult the collections, to have them issued with an admission card.

The problem of access to the botanical garden became a recurring issue. On September 27, 1881, Eugenius passed on a pharmacist's repeated request to use the botanical collections: "I would appreciate an answer early tomorrow"[3]. Eugenius' notes were often very brief, even reduced to instructions, which could not help attracting the good graces of his interlocutor.

Eugenius kept a watchful eye on Didrichsen. For example, on November 7, 1879, he wrote to Didrichsen that he found it "inappropriate to propose as the subject of a dissertation for a prize a subject already studied by one of your students". Would Didrichsen have wanted to favor him, or was this just carelessness on the part of a disorganized person? Two days later, Eugenius sent him a long letter of almost four pages in which he summarized the reasons for his "bad feelings" towards him. They were purely professional, of course.

Was Didrichsen's questionable support of students a regular practice? It is hard to say. However, there is another example. Letters from March 1879 – not addressed to Eugenius – show that Emil Christian Hansen's request to do his doctorate without having validated a previous university course – for economic reasons, he explained – was supported by Didrichsen.

---

3 Letters held by the Uppsala University Library. See: https://www.alvin-portal.org/.

These practices, which were commonplace until the middle of the 19th century, including in other European countries, were now condemned, especially as the University of Copenhagen offered a complete curriculum.

Was Didrichsen touched by the fact that Hansen had first been a journeyman house painter, then a volunteer assistant to Japetus Steenstrup? Didrichsen himself humbly began his career as a librarian at the botanical garden. Hansen's career was to be a brilliant one. From 1879, he headed the physiology department of the Carlsberg industrial laboratory, having successfully defended his thesis in 1877. He worked on alcoholic fermentation, discovering a new technique in 1883. Of the same generation as Eugenius, he was born in Ribe and has his own memorial there.

Logically, after Eugenius' frontal clarification in the long letter of November 1879 already mentioned, in which he recounted the history of grievances accumulated since the end of the 1860s, their relationship did not improve. Their differences were even exposed in the public arena, through *Berlingske Tidende* (*Mr. Berling's News*), a major conservative daily paper close to the government, owned by the Berling family[4].

Today, Eugenius Warming's name is attached to the garden he directed for a quarter of a century, while Didrichsen's is not mentioned, even though he preceded him in this role. Admittedly, he found himself in an uncomfortable position. He occupied a position despite the mixed assessment of his qualifications by the faculty and the reserved opinion that was finally issued. He remained isolated within the university, which weakened his ability to plan scientific work. In a way, he lived in the shadow of Eugenius, who was only a *docent* at the university until 1882. He taught advanced students and was in charge of practical work in plant anatomy. He also taught at the Polytechnic School (1874) and the Pharmaceutical School of Copenhagen (1875).

6.4. A pleasant opportunity

At that time, there were no promotion prospects in Copenhagen, either at the university or in the botanical gardens. Hanne, his wife, observed that Eugenius was happy when he left with his botanist's box slung over his shoulder and his lunch. He also appreciated it when she accompanied him.

Overloaded with courses and poorly paid, feeling offended by the episode that saw Didrichsen occupy the coveted university post, he distanced himself, in the

4 Letter from biologist Alfred Peter Carlslund Jørgensen to Didrichsen, November 17, 1880.

literal sense of the word, by moving more than 600 km away. At the beginning of June 1882, he accepted a position as professor of botany at Stockholm University, on a salary of 7,000 kr. His entire family moved three months later. The move provoked political tensions, and the conservative newspaper *Dagbladet* was outraged that Denmark was depriving itself of one of its finest young scientists by failing to offer him conditions worthy of his talents.

The Copenhagen Botanical Society held a party at the shooting range on September 9 to mark his departure. He was able to take full measure of the reputation he enjoyed. Among the 80 attendees were professors and lecturers from the university, polytechnic, business school and agricultural college, alumni and members of the Botanical Society. He declared that he was leaving his homeland for the second time, for quite a long time. But this time, his heart was heavy, as he and his family had been living in Copenhagen for several years. He was very active in the Botanical Society. He published numerous studies in his bulletin and attended its monthly meetings assiduously, until his departure for Sweden. He was still present at all the ordinary meetings held until May 1 (Foreningsmøder 1882, pp. 4–11).

On September 16, he left Copenhagen by boat for Malmö, then on to Stockholm, probably by train. The first Mälmo–Stockholm line was opened in 1864, and a more direct one was inaugurated 10 years later (Garreau 1967).

Based on the college founded in Stockholm in 1878, the aim was to create a higher education establishment. Eugenius gave several public lectures on the new physiological and biological orientations in botany. He was able to share his experience gained in Bonn and Munich.

When he arrived at the botanical institute at Klara Strandgata, everything was yet to be built, and that did not displease him. Every cloud had a silver lining! The botanical room was empty. As usual, he worked tirelessly. Hanne was delighted that he had a secluded room in the house, so their five children did not bother him. The youngest, Louise, was born in Stockholm in 1884.

Professor Warming soon made a name for himself. At a meeting of the Copenhagen Botanical Society on May 24, 1884, Eugenius – who was about to embark on a mission to Greenland – announced that five rooms were being fitted out for study, collections and for his assistant. His audience included teachers, botanists and medical students. A few women also attended the school. His program included teaching anatomy for one term, followed by exercises to help students become self-sufficient and literate. Without a botanical garden or fresh plants, he worked with samples preserved in wine spirit (Foreningsmøder 1884, p. 106).

A year later, he left a Swedish and a Danish herbarium, the latter largely composed by a lecturing botanist, Hans Mortensen, a specialist in Danish flora, particularly that of the island of Seeland (*Sjælland*). Collections preserved in glass jars – 2,000 specimens in three cabinets – were built up, modeled on those in the Copenhagen Museum. Eugenius, convinced that botany could no longer be practiced without quality equipment, acquired seven modern microscopes – including two Nachet and a Zeiss – thanks to Professor Veit Wittrock. Curator of botanical collections at the Royal Museum of Natural History, Wittrock was involved in the 1880s in the development of the Bergius Garden (named after its founder) in its present location in Frescati. It is part of the Royal National Park, as part of Stockholm's new urban development plan. On December 2, 1882, Eugenius was elected *rector magnificus* for a two-year term, a title that recognized his eminent position in the young university, in scientific, administrative and educational terms.

Didrichsen's successor in Copenhagen, when he took up his post in February 1886 after completing a series of lectures on vegetal geography in Stockholm, tempers had calmed. The protagonists of his professional and personal conflicts had left the scene, and Didrichsen did not leave only fond memories with those who were still around (Engelstoft and Dahl 1943). Eugenius was also offered the prospect of new scientific expeditions in the territories of the Danish Empire.

The man who cut his teeth in Stockholm had every intention of taking the botanical garden firmly back into his own hands. At last, he no longer had to submit his requests to a superior who was quick to upset him; now everything went through him. But he had not finished with his unfortunate colleague yet.

In letters to Didrichsen on February 18 and February 26, 1887 (perhaps the last), he asked for reports on the garden since 1877. Did he get them? Didrichsen died on March 20. Pugnacious Eugenius requested authorization to search through the deceased's books and papers, which had been deposited with an antiquarian and bookseller, in order to recover what was the property of the botanical garden (letter dated April 13, 1887). It seems that this was not possible. Three months to the day after Didrichsen's death, he made an official request – in a headed letter in which he was mentioned as the garden's director – to lawyer Carl Christian Torp for the return of objects from the deceased's estate, offered at auction, of which some belonged to the botanical garden. Aside from his relationship with Didrichsen, Eugenius would never have thought of plundering the university.

Didrichsen's obituary written by his colleague Petersen (botany) is of rare virulence. The author expressed his difficulty in characterizing his colleague because of the contradictions that existed in him, both as a man and as a scientist, and the lack of correspondence between his results and his abilities. In his opinion, he was discredited by the younger generation of botanists and by his colleagues. His practices and behavior clashed with academic formalism. He did not hesitate to shock people at university meetings or at the botanical society of Copenhagen. To top it all off, he made no secret of his social-democratic sympathies, which, Petersen conceded, were consistent with his concern for the poor and oppressed and his propensity to open his purse to help them (Petersen 1887–1891, pp. 45–47).

Whimsical, fickle, disrespectful of convention, a social democrat and probably not very religious, he was the complete opposite of Eugenius.

7

Back to Copenhagen

7.1. The teacher-researcher

Eugenius was Professor and Director of the Botanical Garden from 1886 until his retirement on December 31, 1910[1]. His office was located at 140 Gothersgade in a building at the south-west corner of the garden.

Hanne wrote that her husband worked very hard and never seemed to tire, so passionate was he about his new responsibilities. He stirred up dust and cleared out the museum attic with bundles of collections, bringing them out of oblivion, including the seaweed herbarium of Hans Christian Lyngbye, visitor to the Faroe Islands, and the plants from the first Galathea expedition (1845–1847). In a letter to his daughter Louise in March 1906, he wrote: “Life is so rich in the joys of work, and I want to enjoy it a little longer”. He was about to turn 65, in a country where life expectancy at birth was 57, one of the highest in Europe, on the eve of World War I (Prytz 1984, p. 87, p. 140; Vallin 1989, p. 37).

Thanks to the experience he had acquired, and the authority, tenacity and energy that characterized him, in 1888 he obtained a grant from the Ministry that enabled him to direct the construction of a botanical laboratory, which was put into service two years later. He and his family moved into the building. In mid-September, Hanne wrote to her brother to say that their home was very spacious, and that he even had a guest room. Very present, she often accompanied her husband on his morning walks, sometimes on botanical excursions. As the children grew up, she took a growing interest in his work and helped him with it. Together with her eldest daughter Marie, she saw that Eugenius was never happier than when he was out in the field with his precious camera, or in his laboratory or office writing.

---

1 According to Prytz (1984, p. 149), in an undated letter from his wife to his daughter Louise, Warming announced to the faculty that he would be retiring on January 1, 1912.

Figure 7.1. *Botanical laboratory, 140 Gothersgade, seen from Copenhagen Botanical Garden (public domain via Wikimedia Commons). For a color version of this figure, see www.iste.co.uk/matagne/birth.zip*

On an excursion organized for botanists and foresters in the Hørsholm region (25 km north of Copenhagen) by her father in 1894, Marie met her future husband Carl Vilhelm Prytz, professor of forestry at the College of Agriculture. He sent her a bouquet of roses on October 2, her 22nd birthday, and they were married on March 15, 1895 (Prytz 1984, p. 87, p. 99).

Eugenius was finally free to take his students on regular visits to the botanical garden and heated greenhouses. He introduced them to exotic species in the palm house, arborescent species and orchids from the tropics, aquatic and palustrine plants from warm countries, Japan, the Mediterranean, North America – which required shelter during the winter – native and alpine plants remaining outdoors all year round, as well as open-air trees and shrubs. Students could study the 2,350 species of herbaceous perennials, plus biennials, annuals and medicinal plants lined up in the flowerbeds, but only a maximum of 900 native species could be seen in the garden, which covered a relatively small area of 2,800 m² (Jacobsen and Rothe 1879).

He felt it necessary to get his students out of the city to study Danish flora and vegetation in greater depth. He took them on short excursions on foot around Copenhagen. But he found this insufficient. He demanded and obtained funds from the government to organize excursions lasting several days, one each year. From 1893 onwards, he alternated the West Jutland he knew so well with the islands of Bornholm (e.g. July 1901) and Seeland in the Baltic Sea (Warming 1901). He drew comparisons between different sites on the eastern and western coasts of Denmark, mentioning the difficulty of access, linked in particular to the tides, which made paths impassable. He then alluded to his native island, which he visited for the first time in 1889.

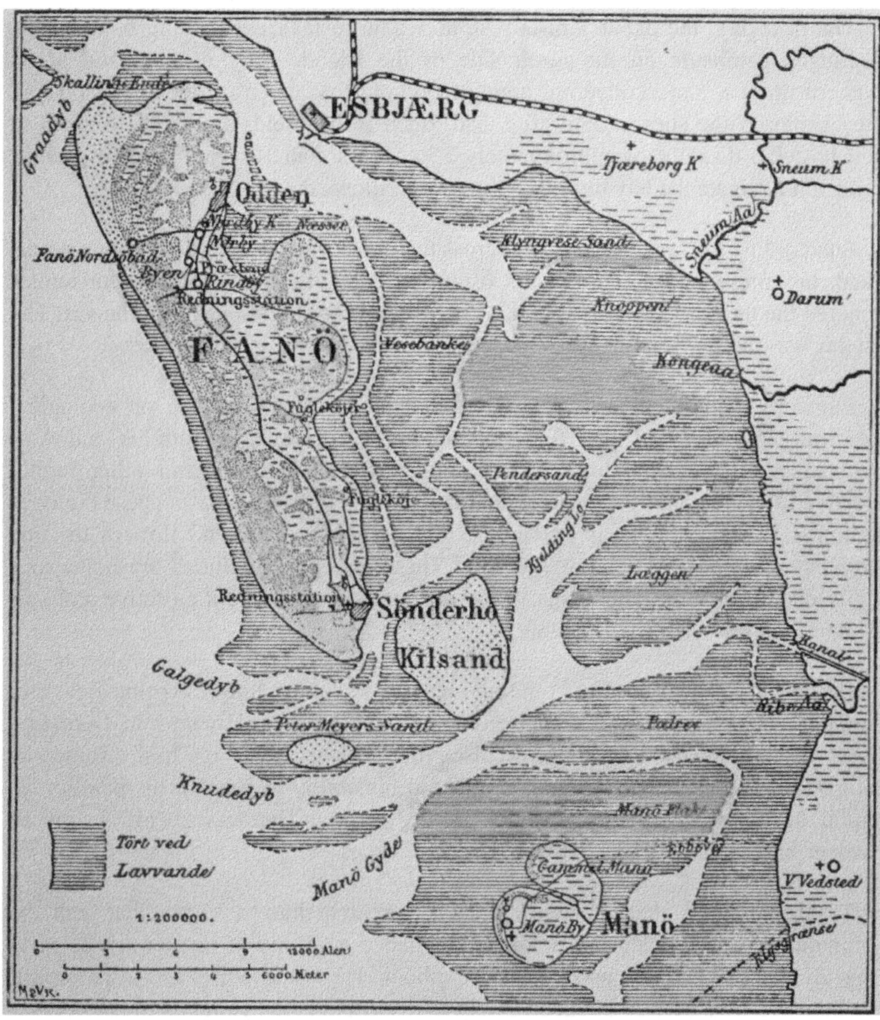

Figure 7.2. *Map of Fanø and Manø (Mandø), with the Wadden Sea (Warming 1906, p. 113)*

On Friday, July 14, 1893, he was on the island of Fanø, one of the largest in North Frisia, between the island of Mandø and the small peninsula of Skallingen, northwest of the port of Esbjerg, with some 20 students of natural history and geography. He presented this excursion as an illustration of Raunkiær's thesis on the plant formations of the west coast (Warming 1893b). Raunkiær, who took part in the outing, identified six: forest, heath, bog, meadow, freshwater and dune formations (Raunkiær 1889; Warming 1894).

The next day, the excursionists rose at 5 a.m. to take two fishing boats to the Skallingen peninsula on the north side of the island. They disembarked at its southern tip at 8 a.m., collecting luggage brought to them from the Blaavand inn. They explored the surrounding area, had lunch in the field, then walked north to their hostel, a distance of approximately 5 km. Less than 10 years ago, large ships could still navigate the bay but silting no longer allows this.

Sunday July 16 saw them set off across the dunes and along the beach. Eugenius noted the presence of xerophytic (drought-adapted) plants. The excursionists explored the dunes, sandy fields and large depressions often flooded by the sea. The last day was disrupted by heavy rain, so the planned outing was cancelled.

Eugenius mentioned these few days in the field in his treatise on ecological botanical geography of 1895. He often used his notes, which fed into his studies of the adaptation of plants to dune environments, salt marshes and other highly selective foreshore and salt meadow habitats on the North Sea coast. These biotopes are found in low-lying areas on clay bottoms, where the ebb and flow of the tide leave emerged areas of land at low tide. These extreme conditions are home to a highly distinctive flora and fauna. His work on the marsh is authoritative and still referenced in the scientific literature.

He invested a great deal of time and energy in his teaching, favoring field and laboratory work, articulated with lectures and conferences. His stay in Germany, where pedagogy was considered a discipline, left its mark on him. Numerous testimonials from his contemporaries attest to his ability to pass on his passion for natural history and botany. Attentive to his students, he was always happy to discover in some of them an aptitude for the sciences.

Teachers know that structuring their research into a form that can be communicated to an educated or trainee audience is a kind of test, and it helps to draw up plans that can be used to outline a book. He would write, and often repeat, that his 1895 treatise was intended solely to disseminate the contents of his courses and lectures to students and his Danish public.

The published versions of his excursions reveal a number of interesting approaches and positions. They foreshadow themes and even chapters of his future treatise. For example, a species from the marshes of the North Sea coast suggests reflections on its morphological and anatomical peculiarities:

> Glasswort is, in my opinion, one of the strangest plants we have in nature. From its general appearance, it's a desert plant: almost leafless, fleshy, juicy like a cactus. By its internal structure, it is no less a drought plant (Warming 1891, p. 216).

This little marsh plant is still widespread today in salty, waterlogged environments. It can be preserved like pickles, eaten like green beans or added to omelets and salads. From a botanical point of view, the species of the genus *Salicornia* are difficult to distinguish, notably due to the homogeneity of their vegetative and reproductive organs and the variability of the physical conditions of the environment (salinity, more or less prolonged immersion, substrate, etc.) (Lahondère 2004).

Eugenius' discussion of glassworts was all the more important as it led him to question the very notion of drought and the ways in which plants adapt to these conditions. Despite his initial remarks on their morphology and anatomy, which may bring them closer to desert plants, he maintained that glassworts did not appear to be desert or drought plants. He relied on a taxonomy that placed them in the halophyte group, designed to bring together plants observed in water or on damp or flooded soils with high salinity. This group was based on a distinction between physically dry and wet environments, established by the Danish botanist Schouw in the 1820s and adopted by most 19th-century geobotanists.

Eugenius widened the question as he concluded: "Although the bottom of the old marsh is hard and dry, the belt in which sedges [*Carex*] and true halophytes grow is still so rich in moisture, both from the air and the soil, that it cannot suffer from drought" (Warming 1891, p. 228). Thus, despite their morphological and anatomical characteristics, he could not accept that halophytes were subject to drought. He maintained these positions, which were attacked after the publication of his treatise, highlighting certain limitations of his ecology, as we will see.

Another question, also concerning vegetation observed during excursions with his students, gave rise to a particularly innovative chapter in his 1895 treatise on the succession of vegetation in the field. In salt marshes, along the brackish and fresh waters of fjords, he identified belts of vegetation – from seaweed to grass meadows – according to their distance from the sea. He explained how certain plants conquer territory: migrations, progressive settlement, adaptations of certain species at the expense of others less well adapted. In the same way, he studied how dunes are formed and colonized by a succession of psammophilous (sand-loving) species. He pointed out that the history of dune evolution provided an example of the struggle between different forms of vegetation, allowing us to follow "the ruin of one, the victory of another", until *Calluna vulgaris* (a heather) takes possession of the soil (Warming 1892a, pp. 167–168). He was similarly interested in the shores of Lake Skarridsø (Seeland Island), where he studied the succession of plant communities (Warming 1897).

His interest in shores and islands, in a country that boasts several hundred of them, over 7,000 km of coastline and more than 1,000 lakes, was not neutral. For a geobotanist, geographical islands – or inverted islands such as lakes – as well as coastlines and foreshores, offer environments that exert strong selective pressure. This makes it easier to detect adaptive changes. Over short distances, several plant formations and communities can be observed with transitions. Environmental factors are very marked and identifiable in the field (prevailing winds, clayey, sandy or flooded soils, slope exposure, etc.). In a way, for students, these environments are pedagogical.

Eugenius supported the research of a number of doctoral students – even though he sometimes fell out with some of them afterwards – some of whom, like Raunkiær, were great scientists. Another example is Wilhelm Johannsen, who contributed to two editions of a botanical textbook, Frederik Børgesen, who sailed to Venezuela and the West Indies with Eugenius in 1891–1892 and became a specialist in marine algae, Olaf Paulsen, who learned botany in Copenhagen and became a specialist in Central Asian flora.

7.2. The successful author

The first edition of the *Handbook of Systematic Botany* published in 1878 by Eugenius was so successful that it was translated into German, Russian and English, and revised and expanded several times (Warming 1878b, the English edition dates from 1895). The latest edition dates from 1933. Several generations of students and teachers have had their hands on it. This voluminous, illustrated work deals successively with all the major taxa, with the different classes and families they contain. Each group is presented morphologically, anatomically and organogenetically. The 1895 English version is 620 pages long and has an appendix on plant classification by translator Potter, Professor of Botany at Durham University (a county in northeast England). In it, he offers a historical reading of classifications from artificial systems to the most recent natural classifications, and attempts to take account of the phylogenetic links between taxa..

In 1880, Eugenius also wrote *Den almindelige Botanik: En Lærebog, nærmest til Brug for Studerende og Lærere* (General botany: a manual of morphology, anatomy and physiology of plants for the use of students and teachers) also illustrated with numerous microscope drawings, such as mitosis figures. It was translated into Swedish and German. Several editions were published up to 1909. In the second edition of 1886, the section on physiology was redesigned, and an independent chapter on development and metamorphosis was added (Warming 1880b).

He was also interested in elementary and secondary education. In 1900, he published a botany textbook for schools, colleges and seminaries entitled *Plant Life – A Text-book of Botany for Schools and Colleges*, which went through six Danish editions until 1920, the last three with the collaboration of Raunkiær and one with Johannes Boye Petersen, another of his former doctoral students who had become a specialist in Icelandic algae. The manual was translated into Russian in 1904, Dutch in 1905, English in 1911 and reprinted until 1919 (Warming 1900a). It emphasizes the importance of biology and physiology in teaching, and takes a stand on evolutionary theories.

Other successes include a collection of 20 botanical wall boards for school use, widely used in Denmark and neighboring countries, created in association with Vilhelm Balslev[2]. A native of Ribe, Balslev held a degree in theology, and had taken additional examinations in natural history and geography. Passionate about teaching natural history, he was trained in botany by Eugenius and published botany and zoology textbooks for primary and secondary schools with Kristen Simonsen, who promoted an experimental approach to teaching natural history, and Andersen. With the latter, Balslev also published wall plates of insects. Himself a teacher, Balslev wrote a manual of practical aids for novice teachers with no formal training who wished to teach their pupils about natural history (foreword dated July 1900). His aim was to promote methods that would arouse children's interest in nature, and in observing it in the environment close to their school (Balslev 1900). During the preparation of the school law of April 24, 1903 (which founded the *folkeskole*, for pupils aged 7–15, and the *gymnasium*, the equivalent of a high school), Balslev took part in drawing up the teaching plan for natural history.

2 Examples: drawing plate of chestnut and elm branches and leaves, 1902 (70 cm × 94 cm), archived as P 288; botanical description "Our cereals", 1902 (100 cm × 70 cm), P 238, etc. In the *Sønderjysk Skolemuseum* archives (South Jutland School Museum, Vejle).

Conclusion to Part 2

From the moment he entered university as a student, Eugenius gave the impression of knowing what he wanted. The late 1860s ushered in a period during which he made a name for himself, beyond Denmark's borders, as a botanist, systematist, flora specialist and specialist in plant morphology and anatomy. He was also recognized as a teacher. He taught at a number of institutions and universities, and made his mark in Stockholm, where everything had yet to be built. His handbooks met with great success in Europe among teachers, students and naturalists. The academies and academic societies promoted his work to a public of amateurs and professionals.

Like the naturalists of the time, he belonged to several academic societies. His choices were strategic. He favored the most prominent national and international structures. He was a member of the Royal Danish Academy of Sciences and Letters (where he was active from the early 1870s until his death), the Royal Swedish Academy of Sciences, the Botanical Society of Copenhagen, the American Academy of Arts and Sciences, the Royal Society of Physiography in Lund (Sweden) and the Bavarian Academy of Sciences (1893). He was an honorary member of the Royal Society of London and of the Botanical Society of Edinburgh (from 1886), and was elected a foreign member of the Linnaean Society of London in 1888. In France, he was a correspondent of the botanical section of the *Académie des sciences* (1904). He was able to obtain financial support from certain societies. For example, he thanked the Natural History Society of Copenhagen for covering the cost of publishing the numerous drawings of his Central Brazilian flora.

From the late 1870s to the early 1890s, family bereavements hit him. Professional tensions and setbacks hardened him, revealing his sense of organization and responsibility, as well as his tenacity and unstinting character. He had an institutional stature and reputation as a top scientist, he published widely and promoted the work of his students, some of whom became colleagues.

A naturalist traveler, he was not an adventurer, as was Aimé Bonpland, a botanist from La Rochelle, Humboldt's companion during their journey to equinoctial America (Foucault 1990). Like Humboldt, for whom he had great admiration, his travels were part of a career and a highly coherent program of scientific research. He had already been thinking about Greenland, at least since 1880 (Warming 1880a). He went there four years later.

PART 3

New Latitudes

8

Arctic and Tropical Missions

Although Denmark's colonial empire went into relative decline in the 19th century, in the Arctic region it retained Iceland, Greenland and the Faroe Islands, visited by Eugenius, as well as the Danish West Indies, where he stopped on his return from Venezuela.

The expedition on the corvette *Galathea* (1845–1847) commissioned by King Christian VIII, who showed great interest in natural history, was an opportunity for botanist Didrik Ferdinand Didrichsen, the colleague Eugenius would revile years later, to make a name for himself. Issues of colonial and commercial geopolitics were central to the long and costly circumnavigation voyage – 3% of the state's annual income – which took the crew of the *Galathea* to the Nicobar Islands in the northeastern Indian Ocean.

At the time, Denmark was at the forefront of European marine scientific research. It established close cooperation between science and the navy.

8.1. Exploring the land of the Greenlanders

On March 18, 1884, Eugenius was invited by the geologist Johannes Frederik Johnstrup on an expedition to Greenland. After consulting his pregnant wife, he left Stockholm for over three months.

He left Copenhagen harbor on May 27, 1884 on the *Fylla*, a gunboat of the Royal Danish Navy under the command of Captain Normann, who was not on his first Arctic mission. Built entirely of wood (1862), the ship was a two-master whose chimney attested to its steam propulsion (500 hp).

Figure 8.1. *Danish Navy schooner Fylla, in Copenhagen harbor in 1888 (Official Danish Navy; public domain via Wikimedia Commons)*

Although this was a geological and geographical exploration, Eugenius was able to embark and take advantage of the opportunity to study botany. The 84-man crew was joined by a small scientific team comprising a botanist, the geologist Holm, Dr. Topsöe, a physicist and mineralogist, and Baron Holmfeld, an artist. Modern equipment and instruments – some European and some American – were taken on board to carry out hydrographic research under the responsibility of the commander and the other ship's officers. In addition, the *Fylla* tracked down American vessels fishing for cod and halibut, in violation of rules established with the US government.

When Eugenius landed in Greenland, the world's second largest island was no longer a *terra incognita* in scientific terms. The geographer Élisée Reclus in his *Nouvelle Géographie universelle*, noted that from the 1820s onwards, Denmark undertook a methodical and intensive survey of the Greenland coast, first in the west, then along the northern coastline. However, by the end of the 19th century, the northeast coast remained unrecognized (Reclus 1890, pp. 92–94).

Denmark's colonization of Greenland was reflected in both the economic and scientific spheres by the founding in 1878 of the Commission for the direction of geological and geographical research in Greenland (*Commission for Ledelsen af de geologiske og geographiske Undersøgelser i Grønland*), a Danish-Greenlandic government institution, by geologist and paleontologist Johnstrup and pioneering glaciologist Rink. The following year, Johnstrup founded *Meddelelser om Grønland* (Communications on Greenland), a journal with studies in Danish, German, English and French. A consultative body, the Commission on Greenland was multidisciplinary (human and social sciences, science and technology). It used funds from the lucrative cryolite mining concessions (for aluminum production) to finance expeditions and innovative research, particularly in mineralogy. Johnstrup undertook several exploration voyages (Iceland, Faroe Islands, Greenland) that advanced knowledge of geology and glacial phenomena. The voyage that took Eugenius on the *Fylla* was also carried out for the Geological and Geographical Survey.

After two stopovers, first in Stornoway (on the Isle of Lewis in the Outer Hebrides), then in Reykjavik, the *Fylla* arrived at the end of June in Godthaab, seat of the government of the Danish colony in southern Greenland, which still had 300–400 *Eskimos*, as they were then known. The arrival of smallpox epidemics with the Europeans caused the population to plummet. The expedition's base camp was at Holstenborg (or Hosteinborg, now called Sisimiut, Davis Strait), on the west coast, where the Danes had settled from the early 18th century. The "Apostle of Greenland", missionary Hans Egede, established his church there, and his son Niels had his home there.

The morning of July 1 was devoted to an excursion, then the *Fylla* set sail in the afternoon, arriving at Sukkertoppen (the "sugar loaf", now Maniitsok) at 3 a.m. on July 2. The island was an important center for reindeer skin processing, managed by the state-owned company *Den Kongelige Grønlandske Handel* (Greenlandic Royal Trade), which administered trade for the Kingdom of Denmark. Bad weather thwarted Eugenius' botanical projects. However, he declared that he had visited the most interesting and impressive place he had seen in Greenland, the vast Sermilink fjord and its glacial estuary on the southeast coast. But with the onset of the short summer, his enjoyment was spoiled by the biting insects that preyed on travelers.

On July 6, the *Fylla* set sail in the evening and, after trawling and deep-sea research, returned to its base camp on the 10th. A long expedition took the explorers to islands and fjords. On the west coast, they reached Sarfannguaq (Sarfannguit) on the 11th, after Kerortusok (Qerrortussoq), in the Amerloq fjord, which empties into Davis Strait, where Eugenius and Holm found plant-parasitic fungi. On July 12 and 13, a walk took them into the Itivdleq fjord. The return journey began on the 14th, arriving at Holstenborg on the 16th, where the *Fylla* was due to replenish its coal supplies.

Excursions continued to the island of Disko, as large as Corsica, where a Danish research station was set up in the early 20th century. The explorers stayed for several days: port of Godhavn on July 19 (Qeqertarsuaq), ascent of Skarvefjaeld on the 21st (almost 600 m altitude), then the Lyngmarken glacier (800 m altitude) on the 22nd and 23rd. They left the island on the 24th and sailed in splendid weather between the icebergs of Disko Bay, which were detached from the Jacobshavn glacier (Ilulissat today), which itself comes from the ice sheet that covers 80% of

Greenland. Eugenius admired the "colossal glacier" and the ice sheet in the background, but again complained of "untamable mosquitoes". On the 25th, the crew reached Kristianshåb (now Qasigiannguit, Disko Bay).

Continuing southwards, the *Fylla* approached the island of Egedesminde on the 27th, then Maniitsok again. Eugenius found this area uninteresting, with no plants whatsoever. Understandably, he had no control over the itinerary. Knowing his assertive authoritarian nature, we can assume that he tried to exert pressure to modify it, apparently without much success.

The expedition returned to base camp on August 1. The next day, Eugenius was pleased to find *Linnæa borealis*, a member of the honeysuckle family. The circumboreal species had been discovered for the first time the previous year by Swedish geologist and paleobotanist Alfred Gabriel Nathorst. The early August excursions were more fruitful, with the discovery of species new to Greenland, *Cerastium arvense* (field chickweed) and two *Carex* (sedges). His exploration around Kangerdlugssuaĸ (Kangerlussuaq), at the bottom of a fjord on the west coast, was also interesting. After which, the *Fylla* headed for "home".

On the return journey, the weather was bad. The ship stopped off at Cape Reikjanaes, Iceland's southwesternmost headland, to search for a new island that had appeared at the end of July, without success. It was doubtful whether it existed, or whether it had disappeared as quickly as it had appeared. Another port of call was Kirkwall, in the Orkney archipelago. The gunboat docked in Copenhagen harbor on September 11, 1884 (The Royal Geographical Society 1885).

This expedition went down in the history of the study of plant-parasitic fungi, "charcoals", thanks to the discovery by Warming and Holm of eight new species. Eugenius encouraged Rostrup to study them. The work was continued by Lauritz Kolderup Rosenvinge and Nicolaj Hartz, who collected others in southwest Greenland. At the end of the century, Swedish mycologist Fries, the German botanist Vanhöffen and the Scotsman Simmons added to their collections (Denchev et al. 2020).

For Eugenius, the trip to Greenland represented an important step forward in his program of ecological botanical geography. He now had at his disposal a large collection (2,000 specimens of flowering plants, 133 mosses and 95 lichens) and observations that enabled him to study and compare plant communities of the tropics, northern Europe and the Arctic, and to work on the limits of major vegetation areas (Foreningsmøder 1886, pp. 204–205, 1888, pp. 73–74).

This program continued the following year with new trips.

8.2. Norwegian Lapland

Eugenius set off for Tromsø via Kristiania (or Christiania, which became Oslo in 1925) and Trondhjem, on the shores of a fjord. He used the recently inaugurated (1877) *Trondhjemsbanen* railroad line. The small island town of Tromsø in northern Norway was expanding. Seal hunting intensified, and the port became the gateway to Svalbard – Norway's northernmost archipelago – for polar expeditions.

From June 24 to 29, 1885, Eugenius made daily excursions on the island itself and to the mountains. He went to Bossekop (Bosekop) in the northeast, then north to Hammerfest, considered the most northerly city in the world. He was particularly interested in coastal flora, especially that of the fjords, whose shores he explored. From July 2 to July 14, he went out every day in the area, heading south to Kåfjord. Following the coastline, it is about 30 km from Bossekop to Talvik, where he stayed. From there, on the 17th and 18th, he explored the western shore of Altafjord. He also studied mountain flora. He returned to Tromsø, where this indefatigable field naturalist botanized the oceanic slope of the Scandinavian Alps – explored by scientists since the 17th century – much richer than the eastern slope, whose climatic conditions are continental. Free to move about, he was able to implement a program of excursions that enabled him to characterize coastal, mountain and oceanic vegetation and sketch out their boundaries.

At the end of July, he returned from Trondhjem via Östersund and Uppsala, with a rich botanical harvest and oysters, which he enjoyed with his family on vacation in Små Dalarö, a popular resort for international guests. It was an island some 50 km from Stockholm, where Eugenius was still in office.

On July 8, 1887, he traveled again by train to Kristiania. There, he met the geologist and glaciologist Rink, who had retired to Norway with his wife in 1882. On the evening of the 9th, botanist and mycologist Emil Rostrup joined them from Trondhjem. The next day, they set off with bryologist Lindberg and two young botanists. The weather was bad, despite it being summer, but they were at altitude. On July 11, they headed up the Foldalen valley via Dovre, 300 km north of Kristiania. An impressionist oil on canvas painted in 1894 by Norwegian artist Harriet Backer depicts this valley in summer. A river, probably the Svone, separated into two arms by an island, winds its way between the mountains, lined by a few houses.

The naturalists spent the night in Kongsvoll, a hamlet on the course of the Svone, where there was an inn on the pilgrims' route. The travelers were welcomed by Professor Axel Gudbrand Blytt, Norwegian botanist, and Swedish bryologist Niels Conrad Kindberg, accompanied by their wives and a number of Swedish botanists. They were in good company, among Scandinavian naturalists!

The site came highly recommended, both for the richness of the alpine vegetation and for the comfort of the accommodation. A house specially designed for excursionists had just been completed. Furthermore, full-board accommodation was not at all expensive, and the view was breathtaking: to the east rose the Knutshö, at approximately 1,700 m altitude. It is tempting to dream of a sport–nature tour, in a region where scientific mountain tourism has been developing since the 18th century. At the time Eugenius visited Norway, Norwegian and Swedish associations were promoting this new form of tourism, as in France with the Touring Club and the Club Alpin. The extension of rail networks in Europe accompanied the craze for the mountains.

The group then moved on to the railway station at Hjaerkin (Hjerkinn) – some 10 km from Kongsvoll – from where Rostrup continued on to the Elvedal valley, further south, while Eugenius and the others stopped at Dombås – some 15 km north of Dovres – in the Gudbrand valley, then traveled the following days through the Romsdalen (another valley) to Molde (on the north shore of the Fannefjord). This stay, followed by one in the island village of Mosterhavn, south of Bergen, allowed Eugenius to become acquainted with the coastal flora, which was very different from that of the Alps. He found many plants from Jutland and similar plant formations, but on a much stonier, rockier substrate.

Eugenius' overland journey on the "little Jæder railway" (an allusion to the fact that this is the narrow-gauge railway network), commissioned in 1878 and linking the ports of Stavanger and Egersund, took him to Ogne (Ogna from 1926), where he spent the day. Strolls along the sandy beaches allowed him to admire the spectacle, unknown to him, of dunes mixed with rounded granite cliffs, polished by the sand and devoid of vegetation.

8.3. Equinoctial America

Four years later, he returned to the tropics. His *Lagoa Santa* was ready for publication and he did not return to the village of Lund, who was no longer of this world. However, in his letters, he often expressed the desire to travel to Brazil again. The political situation made it impossible. The Emperor had been overthrown by the army in 1889, and a republic in the hands of an oligarchy of wealthy landowners had been proclaimed. Eugenius worked to obtain the necessary funds for another equinoctial expedition.

On October 21, 1891, he joined a large expedition on a scientific mission until February 14, when he boarded a British liner. His destinations were Barbados, Trinidad (a British colony), Venezuela and the islands of the Danish West Indies (St. Croix, St. Thomas, St. John), visited several times by Danish scientists. Owned by the King of Denmark, the islands were sold to the United States in December 1916, and marine biology research, previously strongly supported by the Carlsberg Foundation, hit a rough patch.

The 1891–1892 expedition was sponsored by the University of Copenhagen and the Danish Ministry of Education. Eugenius was accompanied by Georg Marius Reinald Levinsen, from the Museum of Zoology in Copenhagen, a specialist in colonial aquatic animals (bryozoans), and Lassen, acting assistant. The latter published a study on the flora of the thermal springs at Trincheras (Venezuela). Two young botanists joined them in January 1892, Børgesen and Paulsen, who traveled at their own expense, as well as future fisheries inspector Christian Løfting. He returned to the Danish West Indies as a zoologist in 1895–1896.

It is worth noting that Professor Warming allowed two students identified by him as promising – some of his most loyal pupils and followers – to get their foot in the door. A fitting return! The two botanists went on to publish in the bulletins of the Copenhagen Botanical Society (Videnskabelige Expeditioner 1891, p. 245; Poulsen 2016, p. 197). Eugenius shared his notes and collections with them – just as Lund had done for him – from Venezuela (Puerto Cabello), St. Thomas and Barbados. His Venezuelan plants are now preserved in Copenhagen's Botanical Museum. Børgesen and Paulsen published The Vegetation of the Danish West Indies (1898), using many of the master's observations and drawings. In particular, they referred to environments and formations observed by Eugenius. For example, *Cereus* and *Opuntia* (cacti) in saline soils, the mangroves of Saint John, the lagoons of the port city of Ponce (Puerto Rico).

Børgesen returned to the West Indian islands several times (December 1895 to January 1896, late December 1905 to early April 1906) to study seaweed. Here again, Eugenius provided him with samples he had collected on the islands' shores.

8.4. Teams of specialists

In order to select the specialists he would need, Eugenius took stock of the scientific data available on the Faroe Islands, especially from the late 18th century onwards. He quoted Jens Kristian Svabo, who visited them in the years 1781–1782 – the islands were then administered by the dual Denmark–Norway monarchy – commissioned by the Danish king Christian VII. He left behind an enormous seven-volume manuscript, which remains unpublished. Eugenius consulted him for the botanical part at the Royal Library in Copenhagen (Svabo 1785). Other authors of the Enlightenment and early 19th centuries relied on Svabo's manuscript, a pioneer of Faroese ethnological studies. The naturalist and linguist Nicolaj Mohr (*Om Maaden, hvorpaa Færøenserne, Skotterne og*

Indbyggerne paa Hetland fange Yngelen af Sejen – How the Faroese, Scots and Shetlanders hunt seabirds) and the priest-botanist Jørgen Landt (*Forsøg til en beskrivelse over Færøerne* – A description of the Faroe Islands)

In the 19th century, new knowledge was brought to algology by the priest and botanist Hans Christian Lyngbye. The English naturalist and geologist Sir Walter Calverley Trevelyan increased the inventory of phanerogams (flowering and seed-bearing plants) and cryptogams (algae, lichens, mosses, ferns) to 573 identified species, while Danish traveling naturalist Frederik Christian Raben made his contributions.

In phytogeography, Charles Frédéric Martins posed the first hypotheses on the origins of Faroese flora, following his voyage on the corvette *La Recherche*, which left Le Havre on June 13, 1838 for Norway, Lapland, Spitsbergen and the Faroe Islands. The expedition was organized by the *Commission Scientifique du Nord*, supported by King Louis-Philippe. Marine physician and zoologist Paul Gaimard was at the origin of this undertaking. He had already completed several "circumnavigations" (on the *Uranie*, *L'Astrolabe*) and campaigns in the North Atlantic and Arctic. International in scope, the 1838–1839 expedition brought together French and Scandinavian scientists, and benefited from the advice and instructions of such great scientists as astronomer and physicist François Arago and Alexander von Humboldt (Cassanello 2008).

Martins left a herbarium of Norwegian species, preserved at the University of Strasbourg. His work on the glaciers of Spitzbergen, Switzerland, Norway and Scandinavia in general was of particular interest to geologists, glaciologists, climatologists and meteorologists (Martins 1840, 1843a, 1843b). He also tackled the fields of botany and botanical geography.

Following this review of the scientific literature, Eugenius considered that the floristic and phytogeographic knowledge on the Faroe Islands was not yet consolidated, even if, as far as botany is concerned, Rostrup and a medical student, Feilbergin, published a critical revision of plant lists in 1870 in *Botanisk Tidsskrift* (309 phanerogams, 612 cryptogams). New studies were therefore awaited. The opportunity arose in 1895, when Danish staff officers visited the Faroe Islands to survey and map. The islands were considered to occupy an interesting strategic position between Iceland and Norway.

Eugenius' publication *Botany of the Færöes: Based Upon Danish Investigations* (over a thousand pages from 1901 to 1908) was financially supported by the Carlsberg Foundation, of which he has been a trustee since 1889. The book, published in three parts, was the result of several years' work by teams of specialists (Warming 1901–1903–1908).

The botanists working under his direction covered a wide spectrum of knowledge of the plant world. They were among the best specialists, whose skills he had already been able to appreciate.

– Børgesen, Eugenius' aforementioned pupil, was an algae specialist. He visited the Faroe Islands six times between 1895 and 1902, for a total of six months. He embarked on several boats, seizing the opportunities offered to him: a small steamer, the *Smiril*; a gunboat, *Guldborgsund* (named after a Danish strait), in charge of fisheries control; and a torpedo boat, *Beskytteren*. He explored the coasts and dredged the deep sea. Equipment was provided, such as a small trawl net and a large iron scraper, which made it possible to harvest seaweed that would otherwise have been inaccessible;

– Jónsson, another seaweed specialist, visited the Faroe Islands from October 26 to December 9 1897;

– Pharmacist and botanist Christian Jensen (Jensen 1885), a specialist known in Scandinavia, took on the study of mosses during a stay from May 7 to August 15, 1896;

– Hartz (lichens) and Ostenfeld (phanerogams), who learned botany from Eugenius, stayed on the Faroe Islands from July 15 to September 4, 1897. Ostenfeld had already made short stays in 1895–1896 with the Danish *Ingolf* expedition (named after the ship) during a zoological exploration of the Arctic seas around Iceland and Greenland. In 1903, he spent three weeks visiting the islands of Strömö (the archipelago's main island) and Sydero (the southernmost);

– As for Eugenius, he reserved biological and ecological research for himself. He was in the field from July 15 to August 8, 1897.

The contributions of half a dozen other authors covered the main taxa, according to each author's expertise. For example, Dahlstedt worked on *Hieracia* (hawkweeds) – a group comprising numerous species whose determination is sometimes tricky – sent by Eugenius and Hartz after their return.

Finally, all 18 of the archipelago's rocky volcanic islands were visited. During this journey, the scientists and soldiers took numerous photographs of plants and landscapes.

9

Eugenius' Arctic Botanical Geography

The publication of *Om Grønlands vegetation* (On the vegetation of Greenland) in 1888 represented a major step forward in the fields of floristics and phytogeography. Eugenius demonstrated the value of making comparisons between different northern areas (North America, Iceland, Spitsbergen, Scandinavia, Russia, Siberia, the Alps, etc.) and, like Charles Martins, of identifying their origins.

Eugenius often referred to the rich collection of Jens Vahl, accumulated during long voyages to the west and east coasts of Greenland between 1826 and 1838 (Vahl contributed to *Flora danicae*, 1849). He also relied on the flora of Johan Martin Christian Lange, *Conspectus Florae Groenlandicae*, which, in 1880, represented the most comprehensive floristic and geobotanical study, bringing together scattered knowledge (Holttum 1911–1912, 1922). According to Holttum, all Greenlandic flora was arctic, a claim Eugenius challenged. He went along with the hypothesis of Charles Martins, for whom Greenland was crossed by a dual flow of botanical colonization. There were several centers of creation from which the flora spread and colonized new territories. For the Faroe Islands, Martins' hypothesis was that there were two major flows. The first originated in Europe via England, Scotland and the Shetland Islands or Norway. At the same time, and in the opposite direction, Arctic plants – from Greenland – spread through Iceland, the Faroe and Shetland Islands to the Scottish mountains, where they found a new home (Martins 1859, p. 236).

This issue of the origin of vegetation distribution areas and the causes of their evolution was at the heart of Eugenius' botanical geography program.

9.1. A comparative and historical approach

In Greenland, Eugenius did not focus on rare plants. He was mainly interested in identifying large regions and plant formations, which he characterized in terms of physiognomy and, secondarily, at a floristic level. On this basis, he considered that Greenland had two botanical regions: the southern birch region and the alpine region, covering the rest of the territory not covered by ice. But he believed that this division needed to be discussed and refined when changing scale and making comparisons.

As we have seen, he explored fjords whenever possible – where Lange found the richest vegetation. He also explored islands, mountains and wetlands, due to their richness and particular environmental conditions. They were in some way exacerbated by climatic, geological and edaphic factors. An example of a formation that demonstrates Eugenius' interest in extreme environments is that of rock flora (*fjaeldmarken*), which grows on rocks. In Greenland, it grows on steep terrain. In the absence of soil, vegetation is scattered, dominated by mosses, lichens and a few herbaceous plants. In Norway, this formation is less discontinuous, with thick, soft carpets. Some of the plants in the rock formation are comparable to those of the heaths, with which he was already familiar, but he felt it necessary to make some subdivisions, as the dominant species may differ. For example, stolon plants are rarer in Greenland, while phanerogams in clumps with strong roots or with rosette-shaped leaves are more abundant (Warming 1888).

Wetlands were once again attracting his attention: fresh waters, marshes, ponds and coastlines. They contain few phanerogams, and algae are still poorly understood. In Siberia, this vegetation is richer in grasses and poorer in sedges. In the marshes of Norway, we find many phanerogams that are lacking in Greenland (Warming 1888, p. 138, p. 235). These phytogeographical observations also apply to environments that have been heavily impacted by human activity. "Smoked earth vegetation" (fertilizer inputs) develops around dwellings, where the remains of hunting, blood, excrement and bones have accumulated over the centuries to form fresh green vegetation, to which are added plants introduced by humans, particularly those transported by ships (32 species of phanerogams). He observed the same phenomenon of soil fertilization in areas colonized by birds.

He deplored the lack of data on northern Greenland, at the limit of the eternal ice. However, he did take air temperature readings which suggested that plants thrived in sunny conditions. On June 26, 1884, at Qasigiannguit, he recorded temperatures that provided unprecedented information, of great interest to the German botanist and phytogeographer Oscar Drude. Like Eugenius, he placed his research along physiognomic lines and they worked together to clarify phytogeographical nomenclature (Drude 1884, 1886–1887, 1897). Drude echoed

Eugenius' observation that, under the sun of Greenland's brief summer, the snow melts and the soil can become scorching hot, drying out mosses and lichens, while vascular plants can only survive thanks to adaptations reminiscent of those found in plants that grow in the steppes and deserts of southern lands.

Eugenius' Arctic travels thus provided him with a wealth of data and material which, as was his wont, gave rise to long-term work, the effectiveness of which can be measured by two ambitious, landmark publications: the first dealing with the botany of the Faroe Islands, already mentioned, and the second with Arctic flowering plants.

Botany of the Færöes highlights the floristic and phytogeographical characteristics of the archipelago. The structure of each contribution to the collective work is standardized: first flora, then phytogeography, except for Emil Rostrup's text on fungi, giving rise to considerations and hypotheses on the origin of the flora in terms of migration phenomena (land, sea, air), geology and debates on the impact of glaciations on flora and vegetation, and their distribution between the different islands.

For phanerogams and vascular cryptogams (mainly ferns), many species were common to those of the British Isles, Iceland and Scandinavia. The hypotheses of migration by air or sea and land connections were put forward. For Ostenfeld, of the 277 species of vascular phanerogams and cryptogams, 70 were Arctic, 164 were temperate European and 43 were Atlantic (Warming 1901–1903–1908, p. 111). In Part II, he dealt with marine phytoplankton, in collaboration with Børgesen for lakes. The distribution of the 269 species of diatoms (unicellular microalgae) between the different northern geographical areas led to the same conclusions as to the low specificity of the Faroese flora. Regarding algae, Børgesen's work on physical and biological oceanography was remarkable and heavily referenced (he submitted his thesis on Faroese algae in 1904). In collaboration with Helgi Jónsson, a specialist in Icelandic algae, an essay was written on their geographical distribution in the Arctic and in the northernmost part of the Atlantic.

9.2. The origin of northern flora

Eugenius' botanical study allowed for the expression of divergent opinions among authors. He was not averse to debate and controversy, while respecting scientific rules. Thus, Ostenfeld wrote in *Botany of the Færöes*: "I regard the whole flora as post-glacial […] The chief part of the present flora of the Færöe (I am only speaking of the vascular plants) has migrated across a post-glacial belt of land". Morten Pedersen Porsild, chief scientist of the Danish Arctic station at Disko from

1906, also maintained that most of today's species were transported by means of a post-glacial land communication.

Børgesen did not believe in this terrestrial connection, and alluded to Eugenius' position, which he shared, that in his opinion these terrestrial connections had never existed. The authors enriched the debate by mobilizing other botanists. Børgesen was attacked by Porsild and Simmons, a Swedish botanist and explorer. They felt he overestimated the effect of seaweed transport between islands (Warming 1901–1903–1908, pp. 686–687, 1910, p. 112, p. 118). In general, hypotheses about the role of migrations by birds, sea currents or a terrestrial connection were considered, as were recent introductions of human origin.

Eugenius concluded Part II with a lengthy argument that went beyond botany and phytogeography (Warming 1901–1903–1908, pp. 664–681). He embraced all naturalist fields, including geology and climatology. He referred to numerous authors, whose opinions were divided between those who supported the idea of a belt or land bridge that once linked all the land masses, and those, like him, who did not. With regard to plant migration, he believed that the three most important vectors – over short or long distances – were ocean currents, wind and animals (especially birds). The human factor came later and was not to be overestimated.

In another publication, contemporary with the one on the Faroe Islands, on the history of the Danish plant world since the Ice Age, he took stock of the current state of knowledge based on an up-to-date review of works (85 references):

> During the melting of the ice, in the "Ice Age", the central parts of the Norwegian and Swedish peninsulas were several hundred feet lower than today [subsidence]; southern Sweden was separated from the north by a wide strait through the central kingdom, where the great Swedish lakes are today; and Finland was connected to the Arctic Ocean. The late glacial ice sea, or "Yoldia Sea" [future Baltic Sea], also covered a large part of Vendsyssel [now an island following the flood of 1825]. To the south, Denmark was higher than today, landlocked by Skaane [Skåne, in southern Sweden] and Meklemburg [western Pomerania]; the West Sea islands did not yet exist, but were part of the same continent; short but powerful rivers […] carried meltwater from the retreating ice mass, for example across the West Sea islands. The Fehmarn Strait [between the islands of Lolland and Fehmarn] and the Langelands belt [Baltic Sea islands] to the Kattegat [between Denmark and Sweden] [is] a river system that we find today as ledges in the seabed.

> During the Late Glacial period, movements of the earth's crust caused the land to rise, the Late Glacial uplift; the Central Swedish Strait dried up, and the Baltic Sea, at least its large eastern part, became a vast freshwater lake […], Lake Ancylus [on the site of today's Baltic Sea], named after the freshwater snail *Ancylus fluviatilis*, which probably first flowed over the Middle Sea [intracontinental sea], but later, when the northern part was higher, made its way through the Sound [Sund, strait between Denmark and Sweden] (Warming 1904a, pp. 5–7).

With the "great subsidence of the North Sea", the waters were able to drain and the sea became salty, as indicated by the aquatic paleofauna, followed by climatic oscillations during a so-called "interglacial" period. Eugenius constructed a table summarizing the main probable characteristics of the vegetation, in order to shed light on the history of the plant world in relation to geological conditions, the animal world and the arrival of the first human communities. He admitted that it was still very incomplete, with many questions still to be answered (Warming 1904a, p. 9). The period considered corresponds to the beginning of the Holocene – approximately 11,000 years ago – and continued to the present day. It is widely believed that the Yoldia Sea appeared at the beginning of this interglaciary period, after the sea had invaded the Baltic proglacial lake, representing the first stage in the evolution of the Baltic Sea.

9.3. Ice Age debates and controversies

Eugenius, continuing his analysis, compared the situations of the Faroe Islands and Greenland:

> The flora of the Færöe is Temperate-European and Atlantic with a touch of Arctic, viz. on the higher mountains […] the flora of the Færöe is, moreover, seen from a geological point of view, a young flora: it had no endemic species of vascular plants – except, as mentioned above, among the Hieracia – (see Ostenfeld p. 107) [Warming 1901-1903-1908], and the few, hitherto unknown forms of the Mosses, Freshwater Algae, Freshwater diatoms, and Fungi, that have been met with, will doubtless, on a closer examination, be found also in another countries.
>
> The Færoes thus form a strong contrast to other Atlantic islands, viz. the Azores, Madeira, and the Canaries, which are rich in endemic species, and have a flora which is very old, related to that of the

Tertiary time; this can only be accounted for by the fact that no Glacial Period destroyed the old plant-world of these islands.

The flora of the Færoes, on the other hand, was no doubt utterly destroyed during the Glacial Period, at least as regards the higher organized plants (vascular plants) and it is doubtful whether any of the lower plants survived this on the highest mountain tops. At that time, it is hardly possible that any other plants occurred there than such as we find growing at the present day on the inland ice of Greenland, and on the everlasting snow and ice fields of similar countries or mountains […] The flora is then to be regarded as post-glacial (Warming 1901–1903–1908, p. 660, pp. 662–663).

In other words, the situation was different between the Faroe Islands and Greenland. In the latter case, some species would have survived the glaciation to the present day. So, there is no reason to think that vascular plants in Greenland could not have survived the Ice Age.

This analysis of the origin of Greenland's flora placed him on the side of the proponents of the "refugium theory". It is based on the "*nunatak* hypothesis", according to which species survived the Ice Age on ice-free lands, or glacial islands (*nunatak*). For the present day, Reclus referred to these islands, which in summer are covered with mosses and phanerogams: "It is a mystery of Greenland that these small centers of plant and animal life exist in the midst of infinite snow" (Reclus 1890, p. 102).

Botanists who supported this theory opposed the *tabula rasa* proponents. They believed that today's flora was the result of the post-glacial reconquest of Greenland after the glaciers left the land bare. Such was the case of Alfred Gabriel Nathorst, with whom Eugenius came into conflict on this issue (Nathorst 1892).

Debates about the Ice Ages were all the more heated given that their existence was attested only in the 19th century, hence the interest of scientists in the "Ice Age"[1], even those who were not specialists in the field (Penck and Brückner 1909; Legros 2019). In a study published in 1925, Ostenfeld leaned towards a "middle term". A few species would be able to live in harsher conditions than previously thought – he based this on the discovery of eight species – but this would not be the most important part of the flora, far from it (Ostenfeld and Hansen 1925, p. 37).

1 Geologists Albrecht Reinard Berhnardi, Jean de Charpentier and Charles Lyell, paleontologist William Buckland, zoologist Louis Agassiz, etc. The expression "Ice Age" was coined by German naturalist Venetz Karl Friedrich Schimper (1837).

The "refugium theory" reached its peak in the 1950s. More recent paleoenvironmental studies have challenged it on the basis of several converging clues, notably based on the identification of source regions and migration routes, which lead to a revision of the chronology of the reconquest of Greenland by animals and plants, clearly post-glacial (André 2000).

Eugenius failed to pick up on one isolated clue: the low level of endemism, which he himself observed. However, it bears witness to the recent nature of the island's flora. In fact, species present in limited distribution areas, such as islands, have colonized them thanks to their mobility and over relatively long geological periods. As we have seen, Eugenius himself noted the case of the Atlantic islands (Azores, Madeira, Canary Islands), with their ancient flora and high rate of endemism.

The last glaciation, which affected northern Europe (Scandinavia, parts of Poland, Germany and Great Britain) from the end of the Pleistocene to the beginning of the Holocene, is now known as the Vistulian glaciation, named after the Polish river. The hypothesis of rising sea levels at the end of this period is well established. However, the melting of the ice cap was also accompanied by a rising of the land.

In the third part of *Botany of the Færöes*, published in 1908, Ostenfeld seems to want to end on a note of calm, asserting his intention not to take sides in the controversy. And yet, he does so in Part I. He states that he only wishes to make a contribution to the ecology of Faroese plants, in line with the program created by Eugenius. It was based on an up-to-date international review of work and on his own observations, limited to the summer period. The most important part concerns the study of plant formations and associations with transitions and adaptations to environmental conditions. After having identified the external factors that influence vegetation: climatic factors, importance of wind, edaphic factors, anthropogenics (crops, livestock, accumulation of waste), he dealt with biological characteristics, referring in particular to the biological types of Raunkiær. He drew comparisons between the Faroe Islands, Denmark and the West Indies (Warming 1901–1903–1908, pp. 907–908, pp. 921 *ff*, p. 1002).

Professor Warming's influence on his former student is obvious.

9.4. Arctic flowering plants

"The structure and biology of Arctic flowering plants", published in *Meddelelser om Grønland* and supported by the Carlsberg Foundation, became a classic work on plant morphology and anatomy in relation to the Arctic environment (Warming 1908–1921). The method was the same as for the Faroe Islands. Eugenius mobilized

10 colleagues and students. One of the contributors was a woman, botanist Agnete Seidelin, a freshwater plant specialist and wife of Raunkiær. At the time, she was known for her drawings of plants in several books.

Eugenius made his collections available, supported authors and was committed to answering their questions. For him, "directing" was not an empty word! It was about contributing to knowledge of the structure and biology of Arctic flowering plants, and providing a basis for future botanical observations and studies.

With the foreword, co-written by Ostenfeld, curator of the *Muséum d'Histoire Naturelle* in France – where the materials collected by the Danish expeditions over the past 35 years are deposited – the aims of this ambitious work were spelled out, and the program avenues mapped out. It is clear that Eugenius was the editor. He wanted to study the conditions under which Arctic species spend the winter, their life cycles, the intensity of photosynthetic activity, climatic and edaphic conditions, with particular attention to climate.

He included the elements of a phytogeographical research program – physionomic – initiated by Krause following the Carlsberg Foundation-supported expedition to East Greenland, commanded by Lieutenant Amdrup (1898–1900). Again with the support of the Carlsberg Foundation, Krause undertook botanical and ethnographic studies of the Angmagssalik (Tasiilak) district on Greenland's least-known eastern coast (1901–1902) (Daniëls 1982, p. 10).

At the beginning of the 20th century, Eugenius' interest in bringing botany into the laboratory was noteworthy. In his view, a new phase was beginning. After a period of relatively short expeditions, the time had come to promote long-term physiological research in the laboratory, thanks in particular to the Danish Arctic Station on Disko Island – at the time the world's most northerly scientific research center – set up in 1906 by the Danish botanist Porsild, chief scientist until 1946. One of the station's assets was its location in a relatively floristically rich environment.

9.5. Boundaries and transitions

The question of the origin of floras and their geographical distribution led to the question of their limits and boundaries. Two formations observed in Greenland were studied in depth by Eugenius: birch and heather heathland. They enabled him to test several clues for determining the boundaries and transitions between phytogeographical areas, at finer scales.

9.5.1. *Birch formation*

Birch forests and coppices are found in Greenland's fjords, which penetrate deep into the land. The birches are small and their morphology is highly modified by the conditions of this environment (3–5 m, twisted). The associated herbaceous species, predominantly grasses, were not yet very known. He noted 55 species of European, American and some endemic types in the southern part of Greenland.

Birch was found in Lapland. The Faroe Islands and North America can also be included. Where the cold is more intense, conifers displace the birch. As for Iceland, its attachment to the birch region was only theoretical, due to human exploitation and destruction by domestic animals.

Greenland is poorer in willow than Scandinavia, and lacks many of the herbaceous species that are common there. Eugenius relied on statistical studies of species proportions and possible migrations. Consequently, according to him, only southern Greenland truly belonged to the birch region – as noted by Lange in his 1880 flora – due to the similar nature of its climate with that of Norway and Iceland. His excursions to Lapland confirmed that Norway belonged to the Alpine region. He also referred to *Flora Lapponica* (1812) by Swedish botanist Göran (Georgii) Wahlenberg.

Discussions, which led him to consider the forest formations of the Urals, Siberia and Behring, were aimed more broadly at determining the extent to which Greenland belongs to the European (Scandinavian) phytogeographical area, and thus laying the foundations for a worldwide phytogeographical classification. The focus here was again on climatic conditions, with the exception of manured soils, where vegetation is more dependent on the physical and chemical nature of the soil.

Closer cuts revealed other formations within the birch region: "willow osier-bed (*pilekrat*) and grassy pastures (*urtemark*)". These are found in sheltered, sunny valley bottoms, where the black, fertile humus is colonized by earthworms. *Pilekrat* are rich in moss near rivers. Woody species include alder, birch, willow and juniper, accompanied by herbaceous plants whose floristic composition changes in drier areas and towards the north. Spaces lined with low vegetation, contiguous with the willow orchards, without shrubs or tall grasses, are found in depressions or at altitude. They could be compared to fields of grasses (with annual and perennial species), but they give way to perennials (*stauden*). "I have called them *urtemark*," wrote Eugenius (Warming 1888, p. 11, p. 13, pp. 27–44, pp. 227–228).

His approach led him first to identify the dominant species, those that gave its physiognomy to the plant landscape, and then detailed the floristic composition. In other words, he looked – literally – at the plants, stood up and considered the overall appearance of the landscape. His review of the literature then completed his comparative study, broadening his perspective still further: Iceland's willow orchards were much more European, while in Siberia, certain *tundra* "oases" appeared to be *urtemark*, as in New Zealand (the Russian archipelago of the Barents and Kara Seas). Using this method, he identified regularities in nature, based on physiognomic similarities. This approach was essential as it enabled the validity of the theory to be tested in the field.

9.5.2. *Heather heathland*

A heather-dominated formation, the moor was well known to Eugenius, who had walked it so often in Jutland since he was a teenager.

It covers much of southern Greenland, where it is dominated by small, mostly evergreen shrubs (15 species identified), accompanied by herbaceous plants, mosses and lichens. It is comparable to other European heather heaths. In Greenland, however, it is confined to the flat terrain found on hills and low mountains. The soil is dry, black, sandy, with few earthworms, often gravelly – the bedrock is close to the surface – water runs off or evaporates quickly, and the soil heats up easily under the action of the sun. Dead leaves remain attached to the branches of shrubs for years, slowly turning to dust to form humus. Together with mosses and lichens, this forms peat, as in Jutland.

The drought of heathland plants did not escape Eugenius' eye or his feet, and he had been sensitive to these issues for a long time. In summer, after the abundant humidity caused by melting snow, lichens turned to dust when walked on. He recognized in the species of these moor structures analogous to those of desert plants: leaves curved backwards, with a pinoid structure (shaped like a pine needle), narrow and appressed (applied on the organ that beared them, but without adhesion), smaller than those of the same species in southern latitudes, epidermis with a wax coating, very thick and cutinized, transpiration-limiting devices (closed or few stomata) and pilosity.

Heather heathland is also found in Iceland and Scandinavia, other parts of northern Europe, northeastern Siberia, Lapland and North America, but not in the northernmost part of Greenland (Warming 1888, pp. 41–68). This was another argument for moving the phytogeographic boundary of Europe somewhere in the southern Greenland region.

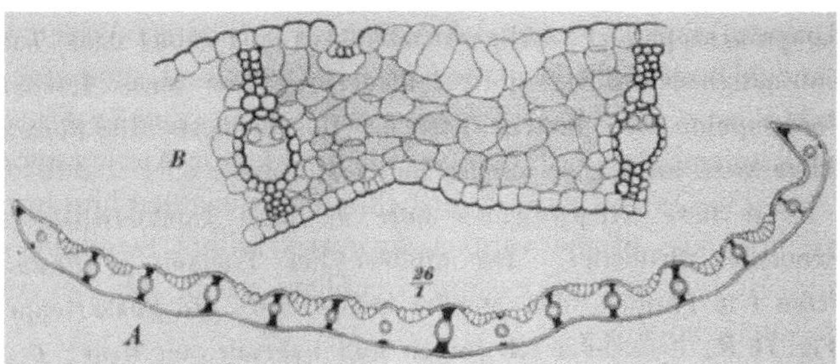

Figure 9.1. *Hierochloa borealis (synonym* Hierochloe odorata, *sweet grass from Tromsø (Norway). Drought-adapted meadow grass. Cross-section of a leaf, with two enlargements (A and B) (Warming 1888, p. 124)*

9.6. European or Arctic flora?

Finally, for Greenland, are there any arguments that allow us to reach a decision? As we have seen, Eugenius put his observations to the test in other areas. In Norway (1885 and 1887), he identified rich alpine vegetation in certain areas, pinpointing the limits of spruce and pine, the way in which birch trees gradually give way to heathland on the approach to lakes and arctic plants on reaching mountain peaks (Foreningsmøder 1888, pp. 73–74).

It echoed the discussions and disagreements between authors on the thorny issue of whether the flora of Greenland was Scandinavian (European) or Arctic. It was linked to the debate on the existence or otherwise of a post-glacial terrestrial connection between Greenland and Iceland. Eugenius once again called upon Drude, who, in 1883, had published a study on the evidence of terrestrial communication between Greenland and Europe. The hypothesis retained – but not settled – was that this communication had broken down after the Ice Age due to the subsidence of the Earth's crust and the rise of the oceans. This would have left only islands (Warming 1888, pp. 178–179, p. 240).

Eugenius' arguments were geological (basaltic nature of Iceland and granitic nature of Greenland), floristic (based on common and endemic species, Greenland's poor flora, leading to the hypothesis of a low level of migration, rare species that were probably present before the Ice Age) and paleoclimatic. He concluded: "Grønland is therefore not a European province, a name that should be given to Iceland and the Færöe".

In the field, he noted that combinations of soil, humidity, light, temperature, exposure, inclination, etc. resulted in a wide variety of possible transitions. He identified 386 species of phanerogams in Greenland, including 15 endemic,

44 eastern (European) and 40 western (America, Asia, Siberia). Widening the focus, he identified several botanical zones: the whole of Greenland, subdivided into South Greenland and West Greenland, Spitzbergen, *Novaïa Zemlia* (New Zealand), Iceland, eastern Lapland, the north coast of Siberia and the Arctic region (Warming 1888, p. 168, pp. 236–237, p. 242). Statistical studies enabled him to compare the relative floristic richness of the various territories within the limits of the data available to him. He assumed that eastern Greenland was poorer, but confessed to a lack of exploration by botanists.

Some data reflect more the wealth or poverty of botanists than that of the flora!

10

The Long-Term Tropics

10.1. Back to the tropics

After publishing his *Lagoa Santa* (1892) in Danish, followed by Danish and German editions of his treatise on ecological botanical geography (1895, 1896), Eugenius revisited his tropical studies in a lecture given at the University of Copenhagen to members of the Association of Scandinavian Naturalists. The Association held nomadic meetings alternately in the three capitals of the Scandinavian kingdoms (Denmark, Sweden, Norway), attracting large audiences.

Eugenius' lecture was translated and published in 1899 by the *Botanical Gazette*, edited by the University of Chicago under the title: "On the vegetation of tropical America" (Warming 1899). Eugenius was on the list of associate editors of the prestigious *Gazette*. The choice of this journal was therefore strategic. The author aimed to penetrate the English-speaking readership interested in ecological botanical geography. Those with access to the Danish and German editions of his treatise were already inspired by his research program. They were the pioneers of ecology at the University of Illinois, the future founders of the Chicago School. With the publication of his lecture in English, he could reach a wider audience.

In the 1890s, American botany was poised to make a spectacular breakthrough in ecology, starting in the Northeast (Acot 1988, pp. 63–64). *The Botanical Bulletin*, whose first issue appeared in November 1875, became the *Botanical Gazette* a year later (the *International Journal of Plant Sciences* from 1992). By 1899, the *Gazette* was a monthly publication featuring European authors (Swiss, German, Italian, English, Swedish, etc.) alongside ecology pioneers from the United States, such as Cowles.

The conference speaker looked back on his stay in Lagoa Santa. In fact, he never stopped working on tropical flora and vegetation, thanks to other trips, numerous exchanges of information and plants with his correspondents, and assiduous reading of published works on these issues.

When he spoke of tropical botany, he always had America in mind. He enriched his comments with the work of other botanists who also knew South America (Chile, Argentina, the West Indies, Paraguay, Patagonia, etc.) and reinforced the relevance of his approach. He also drew comparisons with non-tropical environments. A global approach to the world's vegetation was never far from his mind.

He took stock of the floristic wealth of these regions, illustrated by the 150 km^2 surveyed around Lagoa Santa, which he estimated at 3,000 species. He made comparisons with the flora of Denmark, which had only half that number in its 38,300 km^2. If we add Sweden and Norway, i.e. 772,900 km^2, we arrive at no more than two-thirds of the number of species in Lagoa Santa. Scandinavian naturalists uninitiated in tropical botany were in for a surprise. But what interested him was elsewhere: the search for the causes of this floristic richness and how it had adapted. This quest had occupied him for over 30 years.

For readers of the *Gazette* and the audience of Scandinavian societies, scholars, naturalists, not necessarily up to date with the latest developments in botanical geography and his ideas on the subject, this communication was an opportunity for him to present a didactic synthesis of his scientific contributions and convictions.

He formulated two interrelated questions:

– "What is the cause of this multitude of species in tropical nature?"

– His second question concerned the search for reasons why trees had reached an adapted state, with certain peculiarities linked to the tropical climate: "A very difficult question, of which an explanation is at present an impossibility" (Warming 1899, p. 4).

He did, however, mention physiological causes, but said nothing specific about them other than that all trees in the tropics were similarly equipped. The result was their inability to overcome or supplant trees of different species, contrary to what was observed in northern forests, between the beech and the oak, an example he was very fond of. The most likely scenario was that of a tropical environment in which the trees, over a very long period without any major bioclimatic accidents, had long since reached an adapted state. In this environment, the physical factors which, for him, had the greatest impact were practically optimal: temperature and humidity. Only light remained a limiting factor under the canopy.

The concept of epharmony by French botanist Julien Vesque captures this state of being adapted to a particular biotope. As long as it remains stable, epharmony materializes the mark of the plant's adaptation to its environment. The primary forest is therefore in a state that gives off an impression of equilibrium. Like other temperate botanists, he believes that the optimal environmental conditions prevailing in tropical forests are the most favorable for the production of adaptations.

For plants, the tropics are seen as an "evolutionary Eden" (Raby 2017, p. 29, p. 54).

Eugenius came to a central idea:

> Nevertheless, there is one condition that no doubt has played a conspicuous part in the process; I mean the history of evolution of tropical nature. All characteristics of a vegetation may be classified into two groups of factors, historical and physical. All alterations that the climate, the ground, and the surface of a country are passing through from time to time must express themselves in the vegetable world up to this very day. Consequently this is the case with the inner highland of Brazil where Lagoa Santa is situated (Warming 1899, p. 5).

He began by reviewing the geological history – from the Paleozoic onwards – of what is now South America, citing Lund, the discoverer of "the fossil remains of extinct animal species". He did not mention human remains. He no doubt considered that this communication was not intended to rekindle controversy, which was outside his scope and field of expertise.

Secondly, he attributed the vegetation's characteristics to the exceptional climate, which had spared it from being disrupted by glaciations. Unlike arctic plants, tropical plants have therefore had plenty of time to differentiate and enrich themselves. He assumed that Scandinavian flora must have been much richer than it is today, but that it was decimated by glaciations and has not yet had time to restore itself. But, he predicted, in time – and even without human intervention – it would regain its richness. The variety of tropical nature was also due to "the immense mass of curious ecological adaptations" (Warming 1899, p. 5).

He paid tribute to Darwin:

> It was Darwin's monumental work that led science into new ways. Formerly the botanists especially wished to discover the many unknown genera and species. Now it is life itself they wish to study, those wonderful relations of the varied and often very complicated

> reciprocity of action between living beings, of which a great deal is already known, such as mimicry, the fecundation of plants by insects, the adaptation of epiphytes and other plants to surrounding nature, ant-loving plants, insectivorous plants, parasitism, and other ecological facts (Warming 1899, p. 6).

This tribute was perhaps not unrelated to Darwin's book on insectivorous plants (1875), in which Eugenius is quoted several times. A longitudinal section of a gland from *Drosera rotundifolia*, a carnivorous plant that traps insects, was even reproduced from one of his drawings (Darwin 1875, p. 6).

However, Eugenius did not address the Darwinian concept of natural selection, and did not give rise to a debate on the heredity of acquired traits. He did not even raise the question of the struggle between species. He limited himself to the struggle between species against the constraints of the environment. We would think he would be skirting around the issue, while at the same time pandering to his listeners and, above all, to future English-speaking readers.

Among them, of course, were Darwinians. Darwin's ideas reached Great Britain and crossed the Atlantic. Part of continental Europe, notably France, resisted at least until World War I (Conry 1974). The Danish context will be analyzed below. In this text, Eugenius celebrated Darwin, presented as a great naturalist, with an emphasis found in the English transcription, characteristic of lectures given before an audience of academic society members.

He preferred to develop at length, for Brazil, various climatic, geological and edaphic hypotheses relating to the replacement of the forest by *campos.* We can tell he was at ease. He knew the issue inside out and his work was undisputed. Some British and American authors had yet to contribute their critical analyses. He drew comparisons with other regions of Brazil and South American countries (Venezuela in particular). They confirmed that it was difficult to know which of the two explanatory factors for modern day flora predominated: historical or physical. He based his demonstration on the study of three formations he had studied in tropical America: the forest, the scrub observed in Venezuela and the West Indies, which he equated with the maquis, and finally the savannah (the *campos*), similar to the steppes of southeastern Europe. No doubt he does not underestimate the limits of these comparisons, but he needs to capture the attention of a public and readership more familiar with temperate continental and oceanic formations.

The physiognomies of tropical vegetation are marked above all by two physical factors: temperature and light, which help us to understand the differences between tropical plants and those of northern climates. These two factors naturally have an impact on the phenomenon of plant transpiration, as well as on soil water. Without oversimplifying, Eugenius needed to highlight what would be most relevant for a conference. He described the three formations – forest, scrubland or maquis, savannah – and specified the nature of the climate and the alternation of seasons. He looked back on his recent trips to South America, and gave his personal impressions:

> Hardly any vegetation can be more dis- agreeable and ugly than this scrub of brambles and thorns. Thorny cactus and agave plants, with grayish and brownish hairy leaves, that in the dry season partly fall; everywhere a baking heat and dryness, in some places an immense number of gnats; such is, in short, this scrub which we in northern regions never saw the like of; only in the Mediterranean countries, in the so-called Maquis [in French in the text], do we find anything similar.

And further on:

> And even admitting that the campos of Brazil especially are much richer in flowers than are our meadows, and that the flowers of the campos are much larger and with more gaudy colors than ours, I must say that the wonderful freshness and charm so peculiar to our green, luxuriant, thick, and soft meadows and grasslands I have never seen anywhere, with the exception of a very brief time in the early spring.

Eugenius did not appreciate the dryness of the Mediterranean or certain tropical environments during the dry season. He undoubtedly preferred the mists and rains carried by the ocean winds of Jutland's west coast. More direct in the second extract, he declared that nothing beat the beauty of his own nation's nature, even if the tropics had their charms. We recall that he was seduced by the cool morning air when he arrived on horseback at Lagoa Santa but, he claimed, it was only a "very brief moment".

He concluded his lecture by saying: "the cradle of mankind most likely was in the rich and luxuriant tropical nature". He quoted the great Swedish naturalist Carl von Linné, "habitat inter tropicos, hospitatur extra tropicos", but added that in temperate climates "man has found his most charming home, and just here he develops himself supremely in intelligence, morality, and strength of character" (Linné 1735, p. 24[1]; Warming 1899, p. 14, pp. 16–18).

1 Linné's quotation is truncated: "Homo habitat inter tropicos, vescitur palmis; hospitatur extra tropicos sub novercante cerer carnivorus".

10.2. His favorite family

Eugenius had a strong inclination for species in the "very remarkable" family Podostemaceae, first encountered in the fresh waters of Brazil. Although they are flowering plants, they have the appearance of algae or mosses. Today, we speak of adaptive convergences, which explain the morphological or physiological resemblances between species subjected to the same environmental constraints. Interpreted within an evolutionary theoretical framework, these convergences are evolutionary.

Eugenius was fascinated by the peculiarities of these plants, which he analyzed as adaptations to an aquatic environment in the tropics, where the high water temperature makes oxygen a limiting factor, unlike the cold waters of temperate northern Europe, which contain a greater concentration of this dissolved gas. Sap-carrying vessels are reduced, as are the tissues that support the plant, due to the water's buoyancy. He described long, creeping stems, the formation of haptera (tendrils) on roots, thalloid (thallus-like) roots and stems, and so on. He noted that roots and similar organs were merely anchoring systems. These species could live fixed in fast-flowing freshwater (Warming 1895, p. 451, p. 456).

Although a first volume of his work was published by the Royal Danish Society of Sciences from 1881 (until 1901), he was still sorely lacking in material, as attested by this letter dated August 13, 1880, written in French and addressed from Paris to the great botanist Asa Gray, professor of natural history at Harvard:

> Sir,
>
> Busy with an anatomical and morphological study of the family Podostemaceae, I'm here in Paris for a while to see what the Natural History Museum contains. I see that it's all dried-out stuff, no material preserved in alcohol. Unfortunately, it's so difficult to get good results from herbaria, when it comes to the morphology and anatomy of aquatic plants. That's why I'm taking the liberty of turning to you.
>
> I don't think it would be very difficult to obtain good material from *Podostemon ceratophyllum*, if you could recommend a botanist in America who would be capable of collecting it and would be willing to hand over his crops to me, naturally at my expense. I could pay for his work and expenses, if he doesn't ask for anything extraordinary. But I would need whole plants and not fragments, as they are generally found in herbaria, especially if rhizomes and roots are involved. I already have a few species preserved in alcohol, and I've been promised several more Brazilian ones, but I need a lot of

material; without that, I don't think I'll get anywhere. If you know such a botanist, who would send me material of *Podostemum ceratophyllum*, or of other species, and you would tell me where I can go, I would be very grateful. Or if you have material in alcohol, and could lend it to me for study for a few months, you would be doing me no less a service. Shall I ask you for a few words on the subject? If you'd like to have the work of writing me just two words, please send them to Copenhagen, as I'm leaving Paris in a few days.

Please accept, Sir, the assurance of my highest consideration and esteem.

Dr. Eug. Warming.

He was desperate. The success of his project was at stake. His hopes of finding something interesting at the Paris Museum were dashed. A man of the field, he was always keen to observe the samples himself. Had Gray responded favorably to his request? Possibly, for they had been in touch by letter over several years, and the Harvard botanist had preserved some of Eugenius' gifts in his herbarium.

In the end, he listed approximately 100 species. Today, approximately 180 have been identified. Only one species is non-tropical, found in Quebec (*Podostemum ceratophyllum*).

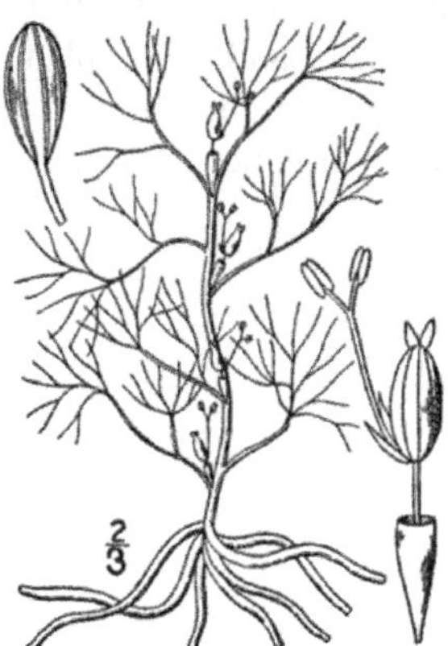

Figure 10.1. Podostemum ceratophyllum *(Britton and Brown 1913, vol. 2, p. 205) (Wikimedia Commons)*

He kept a close eye on any discoveries of new species, such as those made by British botanist Henry Trimen, director of the Royal Botanic Gardens in Ceylon and Möller in Java. His young colleague from the University of Copenhagen, Ernst Johannes Schmidt, during a Danish expedition to the Gulf of Siam in 1899–1900 (Thailand), whose project he supported, discovered the first species known in this region. Eugenius named it *Polypleurum schmidtianum* in his honor (Trimen 1893; Schmidt 1899–1900).

He was identified as a specialist in this taxon by the French botanist and systematist Ernest Henri Baillon as early as the late 1880s, and subsequently by others (Baillon 1888, pp. 262–263; Möller 1899, p. 120, p. 131; Schimper 1903, p. 25, p. 28, p. 236, pp. 817–819; Butters 1917). In the bulletin of the Royal Danish Academy of Sciences and Letters, he regularly published studies and memoirs or information relating to this family, from 1892 to 1915, notably in *Kongelige Danske Videnskabernes Selskabs Skrifter – Naturvidenskabelig og Mathematisk Afdeling* (Royal Danish Academy of Sciences – Department of Natural Sciences and Mathematics) between 1881 and 1901. He was in contact with naturalists who traveled or stayed in tropical regions and sent him samples, such as the Swedish botanist Anders Fredrik Regnell, who lived in Minas Gerais, and Auguste Glaziou, to whom he dedicated a new species. A species was also dedicated to Eugenius in 1914 by Swiss botanists Robert Hippolyte Chodat and Wilhelm Vischer: *Podostemon* (or *Podostemum*) *warmingii.*

The florist and systematist's interest was aroused at Lagoa Santa by the discovery of a little-studied tropical family. The plant geographer who worked all his life on adaptations was already intrigued by the morphological and anatomical changes observed in the aquatic environment (Warming 1896, p. 93, 1902, p. 97).

Conclusion to Part 3

English botanist Richard Eric Holttum, then deputy director of the botanic garden in Singapore, wrote in 1922 that Eugenius' publication on Greenland "is the first ecological study and still the most complete account of the country's vegetation in its entirety", despite some later additions by Morten Pedersen Porsild on various plant families, the origin of Greenland's flora and other less important ecological studies.

The classification of vegetation types proposed by Holttum was partly taken from Eugenius': birch and willow formations (*pilekrat*), herbaceous plants (*urtemark*), heather (*lynghede*), open vegetation (*fjaeldmark*), freshwater marshes and ponds, coastal vegetation and nitrophilous vegetation around dwellings.

Holttum considered the formation of heather moorland to be a climax (except in the Far North), scrub as a postclimax and *fjaeldmark* as a preclimax stabilized due to poor conditions. With some exceptions, the vegetation could be seen as representing stages in the succession of the heath (Holttum 1922, p. 87, p. 91).

These analyses were part of the ecological succession movement which was developing in the United States and England. The concept of climax begins to take hold. The Finnish botanist Ragnar Hult introduced the term and produced a study of ecological succession in his 1881 doctoral thesis ("Attempt at an analytical treatment of plant communities"). He observed, for northern Finland, that pioneer plant communities in relatively large numbers gradually gave way to a smaller number of relatively stable communities, in the absence of disturbance (Hult 1881; Cowles 1911).

Eugenius' study of Greenland was published four years before his study of Lagoa Santa, and seven years before his treatise on ecological botanical geography. "Warming between 1886 and 1891 formulated a new view of how plant distribution was to be studied", wrote Coleman (1986, p. 186).

While he left a lasting scientific imprint on Arctic ecology, he also left a cartographic mark on Greenland. Humboldt had his glacier, *Cape Warming* to the east and *Warming Land*, a peninsula to the far north bounded by two fjords: St George and Sherard Osborn. Warming Land was not far from those of Swedish botanist Thorild Wulff and Danish engineer Marius Ib Nyeboe.

In the 1930s, he was still working on the definition of the great botanical regions. Initially, the zonation of the globe was based on that determined by Drude in the 1880s and 1890s: "By vegetation zone we mean that part of the earth's surface occupied by analogous vegetative forms, characterized by equal periods of activity and presenting the same adaptations and the same protective dispositions" (Drude 1897, p. 58). These groups were biological, not systematic. Charles Flahault, a phytogeographer from Montpellier, wrote:

> Our lithosphere can thus be divided into slices more or less parallel to the equator, into zones clearly characterized by both climate and vegetation. [...] The climatic causes combine in different ways in each fundamental zone to produce, in short, very different climates, and to allow the development of vegetation of very varied appearance and composition (Flahault 1900, pp. 430–431).

For Swiss botanist Josias Braun-Blanquet, founder of the *Station internationale de géobotanique méditerranéenne et alpine* (SIGMA) and of phytosociology:

> The Euro-Siberian-Boreo-American region, the largest on the globe, embraces a large part of the boreal hemisphere, from the Atlantic coasts of Europe through Eurasia and boreal America to the Atlantic shores of Canada and the USA. It is bordered to the south by the Mediterranean, Arab-Caspian, Sino-Subtropical, evergreen regions of the United States (Braun-Blanquet 1929–1930, p. 1).

The Braun-Blanquet zonation took into account deciduous and coniferous forests, dwarf shrub heaths, grass and Cyperaceae meadows, and moss and lichen carpets.

Finally, from a methodological point of view, Eugenius' approach to stimulating and directing work was interesting. It could be described as "modern", since it brought together both established researchers and students whose specialties covered a wide field of study. As Director of Research and Publications, he set the broad guidelines for the work, which was part of a vast Danish scientific program, and obtained funding to send certain authors on missions: for example, Olaf Galløe in 1913, to study lichens.

He took part in the expeditions himself until the late 1890s, when, as he grew older, he stopped planning long voyages across seas and oceans. In 1909, however, he crossed southern Europe on his way to Tunisia and the Sahara. He returned with many photographs. He did not travel to Iceland, although he did edit *The Botany of Iceland*, a book co-edited with Kolderup Rosenvinge, published from 1912 (Kolderup Rosenvinge and Warming 1912–1918, 1918–1920, 1930–1945), and continued by Grøntved, Paulsen and Sørensen (1930–1945, 1942, 1949).

PART 4

Professor Warming's Ecology

11

1895–1896: Ecological Botanical Geography

When the first Danish version (1895) and then the German translation (1896) of his treatise on ecological botanical geography – which was identical to the first – were published, Professor Warming was living the life of an academic, working and writing a lot.

He was one of 60 teachers at Denmark's only university. Since his stay in Lagoa Santa, he had enriched his experiences as a traveling botanist and structured his program of ecological botanical geography.

He had many important publications to his credit. He selected 19 from the bibliography of the *Lehrbuch der ökologischen Pflanzengeographie* (Manual of Ecological Botanical Geography) of 1896.

11.1. The story of a new word

In the Danish edition, *økologiske* refers in a note to the Austrian Haans Reiter, the first to use the word in a work on botanical geography (1885), who cited Haeckel, the inventor of the neologism, and followed the physiognomic tradition (Reiter 1885, p. 4 *ff*). However, his classification of growth forms – one of many – fell far short of Warming's, and his theoretical scope was weak.

It should be noted that the expression "human ecology" was first used in 1892 by Ellen Henrietta Swallow Richards, the first woman to graduate from the Massachusetts Institute of Technology (bachelor's degree in chemistry, 1873). She wrote to Haeckel asking permission to use his neologism. However, for a variety of reasons, not least because she claimed – legitimately – the status of a woman scientist in a male-dominated environment, the use of the word ecology in this sense was not adopted at the time (Dyball and Carlsson 2017). Yet *The Boston Globe* headlined on its front page of December 1 1892: “New science. Mrs. Richards names it Oecologie”.

The term *oekologie* is derived from *oïkos*: house, habitat and *logos*: study, science, discourse; *oïkos* derives from the Indo-European *weik*, which designates the social unit superior to the head of the household.

In the 1896 translation, note (1) refers directly to Ernst Haeckel, creator of the neologism, who in 1866 included *oecology* in a note in Volume 1 of *Generelle Morphologie der Organismen* (General Morphology of Organisms) (Haeckel 1866, p. 8). Acknowledging Haeckel’s authorship of the word ecology, Warming quoted Reiter only to point out that he used the term in much the same sense.

Haeckel defined it as follows in Volume 2:

> By *oekologie* we mean the entirety of the science of the relationship of the organism with the environment, including, in a broad sense, all conditions of existence. These are partly organic, partly inorganic. Both organic and inorganic conditions, as we have already shown, are of great importance for the form of organisms, because they force them to adapt to them (Haeckel 1866, p. 286).

The study of the relationships between living beings and their environment is at the heart of ecological science. Haeckel proposed another definition in 1874, linking ecology to the “economy of nature” and to the Darwinian concept of the “struggle for existence”:

> The *oecology* or geographical distribution of organisms, the science of the entire relationships of organisms with the surrounding external world, with the organic and inorganic conditions of existence; what has been called the economy of nature, the mutual relations of all organisms living in the same place, their adaptation to the environment around them, their transformation through the struggle to live, especially the phenomena of parasitism, and so on. Precisely these facts of “nature’s economy” which, in the superficial opinion of the people of the world, seem like wise arrangements made by a creator carrying out a plan, these facts, I say, when seriously discussed, are necessarily the result of mechanical causes (Haeckel 1874, pp. 637–638; Matagne 2009, pp. 11–16).

In *Économie de la nature* (1749), Carl Linnaeus observes and describes, amazed, the evidence of the Sovereign Creator's wisdom. This providentialist vision, with man as its ultimate goal, was inspired by the natural theology that was gaining ground in Europe. He postulated that the wonders of nature link humanity to the Divine. For Haeckel, this economy obeyed earthly causes. The reference to the struggle for life in the 1874 definition reminds us that he was a great defender and promoter of Darwin's evolutionary theory.

While Haeckel was undoubtedly the creator of the neologism and its first definitions, his contribution to the rise of ecology as a scientific discipline is controversial among historians of ecology. Pascal Acot (Haeckel did not play an important role in the constitution of ecology) and Jean-Marc Drouin (he forged the name of a discipline to which he remained a stranger) consider it to be of little significance, unlike Jean-Paul Deléage, who considers Haeckel's role to be more important than that described by historians of ecology, but remains ambiguous (Acot 1988, p. 17; Deléage 1991, p. 64; Drouin 1991, pp. 80–81; Matagne 2009, p. 5).

More recently, Ariane Debourdeau, a researcher in political science and sociology, has developed an argument that tends to restructure Haeckel's role in ecology. He is best known for disseminating Darwin's evolutionary theory, for his taste for neologisms, for his talents as a popularizer and for the fact that some of his writings were taken up by the ideologues of National Socialism, who nonetheless considered Haeckel's worldview incompatible with their own (Debourdeau 2016).

The term ecology became indispensable after Warming's seminal treatise on plant ecology. From then on, it was used by botanists, phytogeographers and the first generation of ecology specialists of the 20th century, in Europe and the United States, who explored and developed its program, launching new avenues of research.

11.2. Synthesis and innovation

According to Warming, his ecological botanical geography was originally written for his students and lecture audiences. However, the German translation enabled him to broaden his readership.

Indeed:

> From 1880 to 1915, Germany was the center of scientific information, due to the importance of its publishers and universities within the international scientific community. It was the dominant model and a veritable training center for many foreign scientists. Any innovation

> had to pass through this center, which acted as a veritable sounding board, and knowledge of the German language was part of the minimum credentials required of any leading scientist. For example, it was the translation of Danish botanist Eugen Warming's seminal work in 1896 by a well-known German publisher that really launched ecology in Europe (Dupuy 1997, pp. 2–3).

Warming dealt with plant communities (*plantesamfund*, *pflanzenvereine*), a concept coined in Lagoa Santa. The Danish word *plantesamfund* can just as easily be translated as "community" or "society".

The author's aim was to link "floristic botanical geography" with "ecological botanical geography":

> The task of botanical geography is to inform us about the distribution of plants on the Earth's surface, and the laws that underlie it. We can consider this distribution from two different points of view, and, consequently, divide this science into two branches, floristic botanical geography and ecological botanical geography […].
>
> Floristic botanical geography has the following tasks. The first and easiest is to compile a "flora", i.e. a list of the species that grow in a large or small area; such lists form the indispensable basis of the subject. The next step is to divide the Earth's surface into floristic natural zones […] according to their affinities, that is, according to the number of species, genera and families common among them. It is also concerned with the organization of a territory into natural divisions, regions, zones, and their characteristics; with the precise study of the limits of the distribution of species, genera and families (their habitat or territory, area), their distribution and frequency in different countries, their endemism, their interrelationships with the floras of islands and continents, with those of mountains and lowlands, etc.
>
> The thoughtful researcher will not be satisfied with simply observing the facts; he will look for the causes. These are partly current (geological, topographical, climatic) and partly historical […].
>
> Ecological botanical geography has other aims. It teaches us how plants and plant communities adjust their forms and behaviors according to the factors that affect them, such as the amount of heat, light, food and water available (1) [note refers to Reiter].

> A quick glance shows that a species is not evenly distributed over a given surface area, but rather in communities with very different physiognomies. The first and simplest task is to determine which species are found in similar locations (stations). Another task, not a difficult one either, is to describe the physiognomy of vegetation and landscape. Many German and Scandinavian researchers have recently been faced with this task.
>
> The next very difficult task is to answer the questions: why do species group together in defined communities? Why does each [community] have a characteristic physiognomy? Which brings us to questions about the economy of plants, their requirements vis-à-vis the environment and their modes of adaptation to external conditions of their internal structures and their physiognomies, which leads to considering their life forms (*livsformerne, lebensformen*) (Warming 1895, pp. 1–2, 1896, pp. 1–3, 1902, Introduction, Chapter 1).

The first section of the 1896 *Lehrbuch* deals with ecological factors across 18 chapters. It is an exhaustive compilation of knowledge accumulated over the course of the 19th century, to which Warming added his own observations on the environmental factors then identified: air, light, temperature, humidity and precipitation, nature and type of soil (chemical and physical properties), action of plants and animals on the soil, orographic factors (related to relief).

The second section (six chapters) deals with "cohabitation and plant communities" (*Das Zusammenleben und die Pflanzenvereine*). The presentation is built from the most intimate to the loosest interactions. Warming was the first to propose this graduated classification. He pointed out that the boundaries between forms of cohabitation are not clearly defined.

After the analytical approach of the first section, it was time to face up to complexity. He took into account human interventions, and dealt with cohabitation with animals and, above all, between plants: parasitism, heloticism – a form of symbiosis characteristic of lichens – mutualism, epiphytes – specifying that endophytes, like certain fungi, share food, but not epiphytes, which only use another plant as a medium – saprophytes – which feed on dead organic matter – lianas. Chapter 5 deals with commensalism and plant communities. Chapter 6 deals with association groups.

These last two chapters are the most innovative in the section.

11.3. Let us eat!

Warming cited the inventor of the concept of commensalism, the Belgian zoologist Pierre-Joseph Van Beneden, for whom "the commensal is simply a table companion". Van Beneden's text – a pleasant read, full of examples – deals successively with commensals, mutualists and parasites. He refers to those "wretches who cannot do without the assistance of their neighbors and who establish themselves as *parasites* in their organs, the others as *commensals* beside their loot" (Van Beneden 1875, p. 3, p. 83). For example, *Pinnotheres pisum* is a species of small crabs hosted by mussels, oysters and other bivalves. They are commensals, as they consume the food remains contained in these shells. It is a facultative relationship, beneficial to the commensal, until food runs out and the crabs start nibbling the animal's mantle, becoming parasites. Like other interactions, it is a question of balance.

Warming warned his readers. Transposing commensalism to the plant world, he was going to use it in a different sense. The following long extract, translated from *Lehrbuch*, shows the originality of his approach. Before describing commensalism between individuals of different species – which is classic – he analyzed commensalism between individuals of the same species:

> If an association of plants is formed only by individuals of the same species, for example red beeches and nothing else, or by heather, or only by *Aera flexuosa* [wavy hairgrass], we have the purest example. The commensals will then have the same requirements for food, light and other living conditions; since each species requires a certain space and the food is almost never enough for all the descendants, a food competition arises between the plants as soon as the place is occupied by the number of individuals that can develop according to the nature of the species. The most disadvantaged and weak individuals are driven out and killed [...].
>
> The fact that the grouping of individuals of the same species into a community may be beneficial to the species as a whole is understandable; it will, it seems, often be able to maintain its existence in a number of ways. For example, the increased possibility of abundant pollination (particularly in anemophiles) and seed ripening, and it is likely that other lesser-known benefits may derive from communal living. But on the other hand, it brings with it a greater risk of serious damage and devastation caused by pests.
>
> The bonds that connect the same individuals to the same habitat are, as has been indicated, first, the same requirements of life, which are

> precisely satisfied on this habitat, so that the species can maintain itself against its rivals. The unmixed natural associations of forest trees are the result of struggles with other species. But there are differences when it comes to how easily a community can grow and become established. Some species are more social than others, i.e. better adapted to form communities. The causes are biological, in that certain species, such as *Phragmites*, *Scirpus lacustris*, *Psamma (Ammophila) arenaria*, *Tussilago farfara* and *Asperula odorata* [reeds, rushes, beachgrass, coltsfoot, fragrant bedstraw], multiply very easily by means of runners; or others, such as *Cirsium arvense* and *Sonchus arvensis* [thistle, sowthistle], produce buds from their roots; or still others produce numerous seeds which are easily dispersed and can remain capable of germinating for a long time, as is the case with *Calluna*, *Picea excelsa* and *Pinus* [heather, spruce, pine]; or even other species such as beech and spruce have the power to withstand the shade or even to suppress other species by the shade they project. A number of species, such as *Pteris aquilina*, *Acorus calamus*, *Lemna minor* and *Hypnum* [eagle fern, aquatic herb, duckweed, moss], which are social and widely distributed, multiply almost exclusively vegetatively, rarely or never producing fruit. On the contrary, some species, for example many Orchids and Umbelliferae, almost always grow individually.
>
> For many species, certain geological conditions have favored their aggregation into pure communities. The forests of northern Europe are composed of few species and do not mix in the same way as those of the tropics, or even those of Austria or other parts of Europe: the reason for this may be that the soil there is geologically very recent, inasmuch as the time that has elapsed since the Ice Age has been too short to allow the immigration of many competing species (Warming 1896, pp. 106–108, 1902, pp. 111–113).

The "purest example" is therefore represented by a community of plants of the same species. Thus, Warming indicates that when requirements are the same (food, light, etc.), competition occurs only when the abundance of individuals of the same species, linked to reproduction, leads to a shortage of space and food: "The most disadvantaged and weak individuals are driven out and killed." In other words, all guests eat peacefully at the same table as long as there is enough room and food. Competition only comes with scarcity and overpopulation. Warming identified the fragility of monospecific communities due to the risk of destruction by pests and parasites.

In the animal world, competition is a relationship that can be intraspecific (e.g. when two males covet the same female) or interspecific (predation, for example). Warming therefore took liberties with the concept of commensalism, which is usually understood as ecological facilitation between different species, in which the commensal is in a relationship with a host. The former benefits without causing harm to the latter. The Warmingian version of commensalism within a plant community constituted of a single species implies that the "table companions" are all beneficiaries, as long as each finds a place and enough food not to harm the other guests. In this case, we cannot even talk about cooperation. As in certain situations in the animal world, attraction to the same watering hole does not mean that the individuals who find themselves there are in a relationship.

When the commensals are of different species, but belong to the same genus, cohabitation also remains peaceful, within the limits already indicated. In tropical forests, which are highly species-diverse, but in which, according to Warming, the trees have the same physiological capacities and are in an environment where the ecological conditions are very favorable (except for light), commensalism is the most widespread type of interaction. Conversely, in other environments, food competition is fierce. He chose the case of aquatic plants: *Pilularia* (aquatic fern), *Isoetes* (species of this genus have great morphological simplicity due to adaptation to the aquatic environment), *Lobelia dort manna* (whose bluish flowers emerge) and *Littorella lacustris* (now used in aquariums). In some cases, the needs of different species may not be the same: what is sought by one is scorned by the other. In such cases, they complement each other and can occupy the same position without competing with each other.

Finally, the success of plants in competition will often depend on fortuitous circumstances, as well as on their biology, morphology and phenology. This is true of the relationship between trees and many plants on the forest floor: mosses, fungi and other saprophytes, ferns, oxalis and other companion plants, all capable of cohabiting. In our undergrowth, certain spring plants have already completed their cycle when the trees cover themselves with leaves and prevent light from reaching the ground. Warming took three herbaceous plants as examples: the snowdrop and two species of corydalis, which flower in spring[1].

"There is commensalism here. Individuals are at the same table, but dine on different dishes [...] or do not receive their food at the same time". He cited Fernando Höck for his 1892 study of beech companion plants, this latter being dominant (*Leitpflanzen*) (Höck 1892; Warming 1896, p. 109, 1902, p. 114).

1 *Galanthus nivalis*, *Corydalis solida* and *C. cava*.

The beech, already cited by Warming as supplanting the oak in the forests of northern Europe, is like a “monarch” that dominates the other species in the community. However – like a decent fellow – it allows companion species to complete their development cycle. Van Beneden chose another analogy: that of honest industrialists and the miserable paupers they make it possible to survive.

11.4. Plant communities

Warming considered communities from the point of view of their diversity and unity. A certain economy giving them their “character”. This is determined by the ecological factors inventoried in the first section of his treatise, or by the fact that ecological forms of life (*Ökologischer Lebensformen*) are brought together. This “character” is linked to the fact that some species are more social than others, i.e. better adapted to forming communities while others “almost always grow individually”.

This notion was proposed as early as the beginning of the 19th century by Humboldt, who considered this phenomenon of social plants much less noticeable under the tropics than in temperate regions. For Warming, the causes of this sociability were biological: faster multiplication, production of a greater number of seeds, which retain their germination capacity for a long time, ability to support shade or to eliminate other species by the shade they cast. He noted similarities with the animal world, notably in terms of food competition, but he noted that in the plant world there is no sense of the common good, as is sometimes the case in human or animal associations.

Two milestones are important in the emergence of this notion of association. After Humboldt, Schouw proposed a nomenclature that allowed the association to be designated by the name of its dominant species, for example *Fagetum*, dominated by *Fagus sylvatica* (the common beech). The Austrian botanist Anton Kerner von Marilaün described the association according to his own physiognomic and floristic characteristics, without first referring to the environment. The association was then considered an entity in its own right, with an observable reality. This paved the way for phytosociology (a term coined by Polish botanist Jósef Konrad Paczoski in 1896). The study of plant sociability forms the basis of the sigmatist school of the 20th century (from SIGMA, *Station internationale de géobotanique méditerranéenne et alpine,* International Station of Mediterranean and Alpine Geobotany) (Roux 2022, p. 31, p. 35, p. 55).

For phytosociologists, field studies allow them to identify species characteristic of associations and companion species. The former are determined on the basis of the criterion of fidelity. These are species located exclusively or more or less within a given association. When they are present, they are in the company of other species to which they are related: these are exclusive, selective (frequent in one association and rare in others), preferential characteristics (present in several associations, but with a preference for one of them) (Braun and Furrer 1913; Lahondère 1997).

Consequently, phytosociology – based on floristic tradition – relies on associations to provide information on the ecological conditions of the environment; conversely, Warming's approach, heir to the physiognomic tradition, consisted first and foremost of studying environmental conditions and their effects on plants, particularly in terms of adaptation. The association is then identified by its dominant species, i.e. those that mark the physiognomy of the landscape.

Identifying the conditions on which the physiognomy of vegetation depends led him to distinguish six elements: dominant life forms, density (number of individuals), vegetation height, color (e.g. brown evergreen heaths, green heaths with summer foliage, flower color), relationship to seasons (resting and vegetation phases, flowering periods, fruiting periods), species life span (duration of aerial parts, annuals, perennials).

Warming introduces Chapter 6 by reminding us (he refers to the general introduction) that species whose requirements are more or less the same, or are related to each other for other reasons, tend to group together. They are often so closely intertwined that no single species predominates. Slight local variations, for example, in soil humidity, can lead to changes in floristic composition, but without affecting the overall physiognomy of the community. He claimed to have observed many such variations in the meadows of his native Jutland and in northern Germany (the former Danish regions of Schleswig-Holstein).

11.5. Hierarchical units

In botanical geography, at the end of the 19th century, there were several classification systems. Far too many, according to the authors responsible for this inflation. Each one proposed their own system, with equivalences that are difficult to establish between them. In this field of research, we note the strong representation of authors belonging to the Germanic cultural area in the broadest sense (Cittadino 1990).

In the introduction to his treatise, Warming mentions his Danish compatriot Schouw, the Swede Wahlenberg for his Lapland flora (1812), and the German botanist Heinrich Gustav Adolf Engler, who demonstrated the influence of geology on the geographical distribution of plants, August Grisebach, Alphonse de Candolle and Oscar Drude. The last three are of particular interest, as they shed light on Warming's path and choices.

After Humboldt, the German Grisebach was at the beginning of this long history of geobotanical nomenclature of which Warming was an heir. In 1838, Grisebach created the concept of "phytogeographical formation" (*pflanzengeographische Formation*). Warming noted that he transformed it into a "vegetation formation" (*Vegetationsformation*) in a later publication (Grisebach 1838, p. 160, p. 166, 1875). It can be used to group plants with a defined physiognomic character (a meadow, a forest, etc.) and contains a geographical dimension. Grisebach distinguished "plant formations" from "plant forms" (*Vegetationsformen*). The former are communities of plants with a defined physiognomic character. The latter refer to the organization of individual plants.

According to Warming, "life forms" and "vegetation forms" (*Lebensformen, Vegetationsformen*) are the units that play the greatest role in ecological botanical geography as they represent the state of the adapted plant or that of the adapted vegetation. In *Oecology of Plants* (1909), he translates *Lebensformen* as life forms (Warming 1895, p. 4, 1896, p. 5, 1902, p. 6; Millan 2016, p. 2).

As early as the 1880s, he praised the richness of Grisebach's system, but found it complicated and imprecise. To simplify it, in his 1895 treatise, he reduced the 60 or so groups identified by Grisebach to 14. Phytogeographic formation had an essentially physiognomic value for Grisebach, but with a floristic dimension linked to a single species (monospecific formation), to a complex of species belonging to the same family (e.g. grasses), or to different species with points in common (particularly in their organization) (Grisebach 1838, p. 160; Warming 1884). The physiognomic and floristic were therefore juxtaposed, and even in conflict in Grisebach's case. In this respect, it could be said that the original sin lies in the botanical geography inaugurated by Humboldt at the turn of the century. However, the context was different. Humboldt's intention was to open up two lines of research for a scientific discipline founded on new foundations, the "geography of plants".

At the very least, the contributors to botanical geography were united by the idea that there were discontinuities visible in the field. Were they physiognomic? Were they floristic? The debate spanned the century.

In the middle of the 19th century, Alphonse de Candolle – son and continuator of his father Augustin Pyramus's work – published a work of botanical geography which summed up the accumulated knowledge in this field. It had a considerable influence on the direction of subsequent work. It posed new questions that put into perspective the influence of climate on the geographical distribution of plants. "Plants only have a dwelling that conforms to the climate in certain circumstances, in certain countries […], this is probably because previous geological and physical circumstances have influenced them". He emphasized that the distribution of plants was intimately linked to their history, to geological factors and the nature of the soil. His interest in the origin of cultivated plants and their domestication was sufficiently rare at the time to be worthy of mention (de Candolle 1855, pp. 11–12, p. 1340). The explanation of certain discontinuities – the fact that species are distributed in disjointed areas – could then be historical.

He identified "physiological groups […] based on the physiological properties of plants with respect to external conditions" (de Candolle 1874, p. 7). He added: "when we pay attention to the way in which plants behave with respect to heat and humidity, we easily recognize five categories that are more or less in line with geographical divisions": the megatherms ("strong heat and much humidity"), the xerophiles (much heat and plants "fond of drought"), the mesotherms (moderate heat and humidity), the microtherms (European plants, "their character is to require little heat in summer and to dread the cold of winter mediocrely"), the hekistotherms (Arctic and Antarctic, "which are content with the least heat"), wrote Alphonse de Candolle (1874, pp. 8–14).

German botanist Oscar Drude, who worked with Grisebach, continued his work along the same lines, but with an emphasis on the floristic dimension. Within the formations, he identified secondary units recognized locally by dominant species (Drude 1884; Nicolson 2013, p. 98). In Warming's treatise, Drude has 11 bibliographical references. He was a key player in the attempt to clarify phytogeographic nomenclature, although he also proposed his own classification system for plant formations, with 14 formations for German forests alone.

11.6. Warming's classification system

Faced with this body of work, which he compiled after Alphonse de Candolle, Warming decided to start afresh, while remaining true to the traditions of botanical geography.

How was the system constructed? It was an interlocking set.

After the plant itself, the smallest level is that of the association identified firstly by physiognomic criteria and secondly by certain floristics. For Warming, plant associations formed a simple accumulation of individuals, with no real cooperation. However, while there was no individuality in the literal sense, as in associations of people and certain animal associations, there was a relationship of mutual dependence and mutual respect between the many members of an association, in relation to each other. They (associations) form certain organized units, of a higher order (Warming 1896, pp. 109–110, 1902, pp. 114–115). We can see that Warming's associations in the plant world were sometimes linked, sometimes not, to notions of cooperation, or even respect, with a certain vagueness regarding the moral dimension they took on.

Figure 11.1. *Associations on a sandy beach and further inland, West Indies (Warming and Graebner 1918, p. 437)*

These organized units formed the category of "association groups" (*die Vereinsklassen*). "The physiognomy is the same, as are the forms of life, but the species can be different" (Warming 1896, pp. 110–112, 1902, pp. 115–118). For example, the associations of cembro pine, fir, spruce, etc. are in the "association group" of evergreen resinous trees. In the field, some groups are continuous, others discontinuous, especially the scrub in the *campos* of Brazil or those of the coastal dunes.

The "association groups" were numerous. He then proposed higher-level units that grouped them according to an ecological factor that he considered decisive for terrestrial plants: water. He proposed four categories: hydrophytes (*Hydrophytenvereine*, living totally or partially submerged), xerophytes (*Xerophytenvereine*, adapted to dry environments), halophytes (*Halophytenvereine*, he cited samphire, tamarisk and species typical of mangroves, such as mangrove trees, adapted to salty environments) and mesophytes (*Mesophytenvereine*, average water requirements, the largest group of land plants). He therefore took up terminology that was already known, but with more precise content and in a substantive form; the distinctions being physiognomics, biological and morphological (Warming 1896, pp. 120–349, 1902, pp. 127–362).

Warming's system was based on criteria intended to be unambiguous, for both theoretical and practical reasons. Indeed, it needed to be able to be used in the field to identify different phytogeographical units, with the same reliability as flora for species identification. While Warming built on existing knowledge and research into the relationships between plants and their environment, he also generated new knowledge and measured the complexity of ecological phenomena. The great British botanist Arthur George Tansley evoked the joy of testing Warming's system in Great Britain, based on the *Lehrbuch* of 1896.

At the end of the 19th century, Warming thus stood at the crossroads between ecological botanical geography and 20th century ecology, whose program lines he identified. Several points underpin this assertion:

– the consideration of commensalism is innovative in Warming's work. Thanks to Warming, the work of Van Beneden was brought to the attention of botanists and the first specialists in plant ecology (Acot 1988, p. 53; Drouin 1991, pp. 135–137; Worster 1992, p. 221; Poreau 2014, pp. 248–249);

– a synthesis between the physiognomic approach of plant formations – resulting from the sum of growth forms – and the "ecological life forms" (*Ökologischer Lebensformen*) of the individuals that constitute them. Warming's classification can be described as epharmonic, based on the state of the plant adapted to a particular biotope. The concept of epharmony by botanist Julien Vesque is based on the hypothesis that among all the organs of a plant, some have a nature that depends solely on their adaptation to the environment (Vesque 1882, p. 9). Warming cited him several times and took up examples and experiments from other authors on adaptations to different environmental factors;

– Warming raises age-old questions renewed by ecology. How can diverse, heterogeneous elements be brought together? Are there discontinuities in nature?

12

Authorized Editions, or Not...

12.1. The problem of the 1902 edition

The German edition of 1896 was hailed by the European scientific community and by the American pioneers of ecology. Warming wrote in the 1909 English edition that, due to time constraints, he had only been able to make "insignificant" changes to the German edition, and had "been compelled to postpone the more important changes he had in mind" (Warming 1909, p. 5).

Warming thanked the translator, the German botanist Emil Friedrich Knoblauch, for having had the courtesy to seek his authorization, *Plantesamfund* having fallen into the public domain, this was not a legal obligation. Indeed, the Bern Convention adopted in 1886 (entered into force on December 4, 1887), which relates to the rights of works and authors, gives the legal means to the latter to control the way in which they are used, by whom and under what conditions. Denmark subscribed to it only on July 1, 1903, after Germany, Belgium and France, who signed the convention on December 5, 1887.

This was not just a detail.

Most historians of ecology refer to the German edition of 1896 to find the content of Warming's original treatise without going through the Danish text. This is why specialists may be surprised to see that, for each of the notes, the pages from the 1896 edition and those from a 1902 edition, generally ignored, are systematically given.

Here is why.

For the German edition of 1902, the translator was the same as in 1896. However, the name of Paul Graebner, professor at the Berlin Botanical Gardens and

Museum, was added to Warming's name. His main contribution was to review the literature and update the bibliography. For this, he selected authors and publications published between 1896 and 1901 (Graebner signed the foreword on December 31, 1901).

This second German edition posed a problem.

The historian of ecology Robert Goodland indicates that it was not authorized by Warming. According to him, Graebner unduly added his name. Environmental and ecological historian Peder Anker disagrees. He believes it is incorrect to claim that Warming suffered from authorship theft. In his view, Graebner collaborated with Warming, and it was therefore reasonable that his name should appear as co-author on the title page (Goodland 1975; Anker 2002, p. 255).

Is it possible that Graebner simply lacked courtesy or even respect for Warming, and did Warming hold a grudge? In the 1909 preface, Warming states: "With this edition [the one from 1902] I had nothing whatever to do" (Warming 1909, p. 5). Admittedly, he makes no mention of Graebner's request for authorization.

Why bother? There was nothing illegal about it.

Warming did not shy away from battles and controversies when they were worthwhile, from his point of view. Generally speaking, he was a shrewd, even astute man, and certainly not an impulsive one. In 1902, he was 60 years old, his reputation was international, and he had no need to assert himself by fomenting a quarrel from which nothing good would probably come, and for which he would have had no legal backing. As for Graebner, a young professor in his thirties, responding to a publisher's request to update the bibliography of Professor Warming's treatise must have been an honor.

A tangible sign of appeasement on Warming's part: in his bibliography of the 1909 English edition, he mentions the 1895 *Plantesamfund*, as well as the two editions of 1896 and 1902. This could also be seen as a strategy on his part, consisting of recognizing the second German edition and reappropriating it intellectually. He may have felt that the German edition of 1902 – objectively of better quality than that of 1896 – would help disseminate his work, with the added bonus of an updated bibliography. Six bibliographical references to Graebner are listed in the 1909 English edition. Warming cultivated the image of a man of conciliation, working for science and recognizing the contributions of others.

Another German edition from 1918, again with Graebner's name as co-author, had a foreword written by Warming (August 1917). He wrote of the 1902 edition: "The publisher entrusted the publication of the second edition to Professor

Graebner; I myself did not participate in this edition". For the 1918 edition – the publication of which was delayed due to World War I – he spoke of their "collaboration", thanks the other contributors by name before concluding: "I must thank Professor Doctor Paul Graebner, who read my entire manuscript and supplied valuable additions and references" (Warming and Graebner 1918, pp. 3–4).

The 1896 *Lehrbuch* was translated into Polish in 1900. Two independent Russian editions appeared in 1901 (Moscow) and 1903 (St. Petersburg). In all, 42 editions were published in three languages between 1902 and 1933. As for Paul Graebner, a specialist in Central European flora working on phytogeography in Germany, he had not finished with the *Lehrbuch*. He reissued it in 1930–1933. It was his last work, and he died on February 6, 1933 at the age of 61.

After these twists and turns through various translations and editions, we can agree with Peder Anker that Warming did not feel cheated in 1902. As a result, the question will be approached from another angle, that of the content of the 1902 edition, by comparing it with that of 1896.

What interests us here is the dissemination of Warming's work.

12.2. New references

The 1896 and 1902 texts are identical, except for a few details. And like some details, these are important ones. They testify to the dynamism of the emerging ecology.

In the 1902 edition, 16 authors published after 1896 had their bibliographies updated, and 42 authors were new. Were these references integrated into the text of the 1902 edition? Not systematically. Of these 42 new authors, only 16 were cited in the text, where their names were inserted alongside those of the 1896 edition, without comment. The other 26 were confined to the bibliography. Graebner's revised bibliography reinforced the use of the terms *Pflanzengeographie* (seven times), *Vegetationformationen* (twice) and *ökologische* (once).

Almost all the authors added to the list published in German. There were two texts in French and five in English. The French texts were minor, but the most important French-speaking authors of ecological botanical geography already appeared in the 1896 edition. The notes on Arctic plants published in English by the Danish botanist Otto Gelert were also of secondary importance. Unfortunately, his career was cut short by tuberculosis, which took his life at the age of 36.

The addition of Scottish botanist Robert Smith is much more interesting. He was considered by Arthur George Tansley as a founder of ecology in Great Britain. He was encouraged in his studies of ecology by Scottish biologist and sociologist Patrick Geddes. Thanks to him, Smith spent a few months in Montpellier (1896–1897), where Flahault taught him his phytogeographical mapping methodology, which he applied in Scotland (Egerton 2018). Two publications from 1898 and 1899 deal with plant associations, the third on pine and birch seed dispersal (Warming 1909, p. 81, p. 140, p. 143, p. 148). He died at the age of 26 in 1900, and his work was in part continued by his brother William Gardner Smith.

The only Americans listed in the bibliography were Roscoe Pound and Frederic Edward Clements, for their study of the phytogeography of Nebraska (Pound and Clements 1898–1900, p. 24; Gimingham 2003). Along with Henry Chandler Cowles (unquoted), were fervent admirers of Warming. In 1905, Cowles wrote him of his eagerness to meet him.

Clements was one of the pioneers of ecology in the United States. He developed his dynamic study of vegetation – a hallmark of American ecology from the beginning of the 20th century – on the last section of the 1896 *Lehrbuch*, devoted to the struggle between plant communities.

12.3. A new theory

The 1902 edition left room for the mutationist theory formulated in 1901 by Dutch botanist Hugo de Vries. The work cited was *Die mutationstheorie. Versuch und beobachtungen über die entstehung von arten im pflanzenreich* (*Mutation Theory. Experiments and Observations on the Origins of Species in the Vegetable Kingdom*) (de Vries 1901, 1903, 1905, 1909).

De Vries, a contemporary of Warming – he was born in 1848 – was a professor of botany at the University of Amsterdam. He was absent from the two previous editions, although he had already published a work in German in 1889: *Intracellurare pangenesis* (translated in 1910 as *Intracellular Pangenesis*). He postulated that the characters inherited by living organisms are particles called "pangenes" (a Darwinian term), distributed throughout the organism. In 1909, the Dane Wilhelm Johannsen used the abbreviated term "gene" to designate a particle inherited from germinal cells (Johannsen 1909).

For de Vries, evolution was essentially due to mutation, with natural selection playing a secondary role, eliminating the most monstrous individuals. Consequently, the formation of new species (speciation) was an abrupt phenomenon. This was an important point. Evolution is not a slow, continuous process, as Lamarck postulated at the beginning of the 19th century, but discontinuous.

This theme, that of mutation, took its place in the text of the 1902 edition, in the last chapter of the last section on "the origin of species", in the form of a new paragraph of almost two pages. It follows the text that ended in 1896 (unchanged in 1902) with a discussion of hypotheses concerning the origin of species: their appearance was due either to external influences that persist, to heredity or to natural selection. For this discussion, in 1896, Warming called upon several authors: August Weismann, Herbert Spencer, Oskar Hertvig, Franz Krasan, George Henslow and Julien Vesque. The first challenged the principle of the inheritance of acquired characters, the second promoted the idea of evolution and the survival of the fittest, and the third worked on fertilization and the malformations of embryos, a field that would become developmental biology. Krasan discussed the monophyletic or polyphyletic origin of species. These concepts led him to question the hypothesis of a common ancestor. Henslow maintained that natural selection played no role in the origin of species. Vesque had already been cited for his concept of epharmony.

This was the heart of science in the making, with the necessary comparison of ideas, theories and hypotheses that are sometimes contradictory.

The new paragraph in the 1902 edition began as follows:

> More recently de Vries has been particularly interested in the appearance of species by "sudden leaps" (Sprungweisen). In his work of great importance he demonstrates, by precise experiments carried out over many years, that certain forms generated (by "mutation") and which are constant, apart from any hybridization, form constant types (Warming 1902, p. 392).

For de Vries, this mechanism explained the origin of species.

The experiments referred to concern a number of wild species of the genus *Oenothera* (evening primrose, commonly known as donkey grass). De Vries observed, in addition to experimental variations, abrupt and discontinuous variations that were likely to give rise to new species. He was convinced of the validity of the concept of "discontinuous variation". In particular, he worked on a species dedicated to Lamarck, *Oenothera lamarckiana*.

It turns out that de Vries was right, but for the wrong reasons. It is well known now that the variants observed were generally aberrant chromosome segregations rather than mutations affecting genes.

The new paragraph from 1902 goes on to praise the work of gardeners, underestimated by scientists who have long noted the great polymorphism within taxa. De Vries developed examples enabling assessment of what is constant and what is variable, the reasons why changes are stabilized – independently of hybridization – and the role of artificial selection. We know that Darwin first observed the practices of artificial selection before drawing up a general law of natural selection. Readers of the 1902 edition can thus discover the impact of a new theory, that of mutation, on ecology. It enriched the debate launched at the end of the 19th century.

If, as Warming repeatedly asserted, he had nothing to do with this 1902 edition, we can assume that this section on mutations was Graebner's own work. Warming did not reject it, and included it in the 1909 English edition, in the same chapter on the origin of species, and partially revised it to take account of more recent publications and his own position in relation to evolutionary theories (Warming 1909, pp. 369–370).

13

1909: *Oecology of Plants* in the Storm

For the revised and expanded English edition, published in 1909, Warming enlisted the help of botanists and phytogeographers. The two translators were Percy Groom and Isaac Bayley Balfour.

The former, trained at Cambridge University, had, like Warming in his youth, spent time in Germany, at the University of Bonn, where botanist Eduard Strasburger – already cited for his work on the origin of ova – attracted many American and English students to his school of botany. Groom befriended a group of assistants, including Strasbourg botanist Andreas Franz Wilhelm Schimper, who had a major influence on his research. In 1898, Groom was invited by Isaac Bayley Balfour, Professor of Botany at the University of Edinburgh, to briefly head the new physiology department of the university's Royal Botanic Garden. He was then appointed to the Chair of Botany at the Royal College of Engineering in Coopers Hill (Surrey), where he trained Indian students for the forestry service. He returned to London and joined the botanical department of Imperial College of Science and Technology in South Kensington in 1908, where he remained until his death in 1931 (Groom 1932).

To revise and update the references, Warming called upon Martin Vahl, 40 years of age, originally a theologian, geographer and botanist, who had been a lecturer at the University of Copenhagen. Contemporaries agree that the updating of the English edition owes much to him.

At the time of publication of *Oecology of Plants*, Professor Warming was known and recognized internationally as the traveling botanist who had published works in floristics, systematics and phytogeography, as the author of the first study of the ecological botanical geography of Brazil and the first treatise on plant ecology.

An English translation was eagerly awaited.

Arthur George Tansley wondered about the mysterious difficulties mentioned by Warming, without going into further detail, which delayed the release of the English edition, as did Henry Chandler Cowles, then a professor at the University of Chicago, who stated that "a peculiar accumulation of misfortunes of one kind or another prevented an English edition" (Cowles 1909, p. 149).

13.1. Warming's point of view

Warming again insisted that *Plantesamfund* was originally intended for those who had attended his lectures at the University of Copenhagen, to whom he assumed a few other readers would be added. He was surprised by the success of his treatise and flattered by the offer of a German translation in 1896. This edition soon sold out. Positive feedback on *Lehrburch* came from all over the Old Continent, and even from the United Kingdom and the United States, in the nascent circle of early ecology specialists, including the birthplace of ecology in Illinois.

Warming had already published major works in English: *On the Vegetation of Tropical America; Botany of the Færöes; Handbook of Systematic Botany; The Structure and Biology of Arctic Plants*.

He presented his 1909 book as "practically a new one" compared to previous editions and thanked the "young phytogeographer Dr M. Vahl, in order that he may deal critically with purely geographical and climatic considerations" (Warming 1909, p. 5). The most important changes related to the growth forms and their classification, which was entirely new and more detailed, as well as to the parties that dealt with adaptations of water and land plants (Warming 1909, section III, p. 96 *ff*).

The original four sections on communities (hydrophytes, xerophytes, halophytes, mesophytes) were now 13 in number (*ibid.*, sections IV to XVI, p. 149 *ff*). They were based on edaphic and climatic distinctions, as can be seen from their summary presentation in the chapter on "adaptations and oecological classifications" (Warming 1909, p. 136). In the amount of time that his work and administrative responsibilities left him, he said he had endeavored, with the help of Vahl, to take into account the contributions of publications published after 1895, including those of Schimper, the German botanist and paleobotanist, Solms-Laubach and Clements (Schimper 1903; Clements 1905; Solms-Laubach 1905).

Schimper published a treatise on plant geography in 1898 (in German, translated into English in 1903) based on physiological principles. Warming's main point was Schimper's fundamental distinction between physical and physiological drought. From Solms-Laubach he took up paleobotany, and Clements the concept of succession, taken from his own work.

The bibliography of *Oecology of Plants* contains 600 authors, almost double that of the 1896 edition. Of the 300 or so new authors, 115 are from the United States. This shows both the importance of the updating work carried out and the spectacular upsurge of North American ecology in a decade. Historian ecologist Pascal Acot explains this phenomenon using three types of factors: institutional (linked to economic needs), natural (wide open spaces are conducive to a dynamic approach to ecology), scientific and technical (a cohort of "state botanists" inventoried the riches of their employing state) (Acot 1988, pp. 63–64). This fosters the image of pragmatism and efficiency attached to the United States.

Eternally dissatisfied, Warming considered, as he had done in 1895, that he had not achieved perfection. He repeated that plant ecology was still in its infancy and that a great deal of research had to be carried out before the foundations of a natural classification of plant communities could be laid, with a clear, coherent discourse.

13.2. Reasoned opinions

Cowles, announcing the 1909 publication, imagined the pleasure his colleagues would take in seeing their work recognized by the father of plant ecology in the new edition (Cowles 1909, p. 152). Warming's admirers praised the innovations, clarity and precision of the 1909 version. Others expressed reservations and formulated sometimes harsh criticism. The analyses of *Oecology of Plants* by two specialists, the American Cowles and the Briton Tansley, are particularly enlightening.

13.2.1. *A critical admirer*

Cowles was very pleased with the work accomplished on growth forms and the new classification. He found the updates remarkable. The question of adaptations, divided between different headings in the Danish and German editions, was brought together in a single section (Warming 1909, section III, p. 96 *ff*), giving it greater coherence.

For him, the key point was the new classification of growth forms. He noted the "fundamental twofold subdivision into land plants and water plants [leading to] thirteen main classes of plant formations" (Cowles 1909, p. 149). He appreciated the clarification of the concept of formation, which had evolved over the course of the 19th century and had lost precision in various authors' work, to the extent that Warming had initially refused to consider plant formation as an ecological unit.

Warming wrote on this subject:

> A formation may then be defined as a community of species, all belonging to definite growth-forms, which have become associated together by definite external (edaphic or climatic) characteristics of the habitat to which they are adapted. Consequently, so long as the external conditions remain the same, or nearly so, a formation appears with a certain determined uniformity and physiognomy, even in different parts of the world, and even when the constituent species are very different and possibly belong to different genera or families.
>
> Therefore, a formation is an expression of certain defined conditions of life and is not concerned with floristic differences.
>
> The majority of growth-forms can by themselves compose formations or can occur as dominant members in a formation. Hence, in subdividing hydrophilous, xerophilous and mesophilous plants, it will be natural to employ the chief types of growth-forms as the prime basis of classification (Warming 1909, pp. 140–141).

Warming distinguished, within formations, the level of associations for which the floristic component becomes primary. "An association is a community of definite floristic composition within a formation; it is, so to speak, a floristic species of a formation which is an oecological genus" (Warming 1909, p. 145).

This was a new proposal.

All naturalists are familiar with genera and species, the basis of the binominal nomenclature established by Linnaeus. A plant or animal is identified by its genus name, for example, *Calystegia* (bindweed), which groups together several species: *sepium, soldanella* (hedge bindweed, sea bindweed), etc. Plants in this genus have common – generic – characteristics, such as being climbers, wrapping around a support, having alternate leaves, etc. The – specific – characteristics of the species are, for example, white flowers for hedge bindweed, pink streaked with white for sea bindweed. For Warming, the association had the same status as a species, while formation had the same status as a genus. Cowles hoped that this taxonomy would become universal.

Warming had also modified certain groupings. Cowles pointed out that grouping heath and heathland plants in the same category (oxylophytes, plants of acid soils) was a step forward, taking into account the close ecological relationship between these plants and those of cold, salty soils, as Schimper had already pointed out[1]. Warming thus incorporated some elements of Schimper's physiological approach, when he distinguished physical and physiological soil dryness, particularly for class twelve conifers (Warming 1909, p. 310).

However, Cowles wondered whether abandoning the old term xerophyte had resolved the difficulties. It was one of the causes of a multiplication in the number of categories – from four to thirteen – which was not without its drawbacks:

> Among the disadvantages of the new arrangement are many instances where unrelated things are placed together and related things are separated. The most conspicuous case of the former is seen in class 12, the conifers. While this group is a floristic unit, and even an ecological unit from the anatomical stand-point, it is far from being a geographical unit of any sort (Cowles 1909, p. 151)[2].

In other words, he criticized Warming for a classification that was no longer unambiguous. One reason for the multiplication of categories was that Warming's ecology – like Schimper's – was phytogeographical. The nomenclatures were therefore dependent on a great heterogeneity of environments.

Tansley solved this problem by proposing the term ecosystem in 1935, which became an ecological unit freed from a geographical dimension (Tansley 1935). A few years later, the American Raymond Lindeman saw the ecosystem as a thermodynamic system that exchanges energy with its living (biotic) and non-living (abiotic). This approach formed the basis of ecosystem theory (Lindeman 1942; Deléage 1991, p. 127 *ff*).

Cowles also criticized Warming's system for causing terminological inflation built on Greek roots: "it seems a pity to have to brush the dust once more from our Greek lexicons", as the vernacular was preferable (Cowles 1909, p. 151). He made the same reproach for the unfortunate and outdated use of *oecology*, whereas *ecology* made it possible to agree with the modern form *economy*, which had replaced *oeconomy*, derived from *oeconomia*. Finally, he deplored the absence of illustrations. He acknowledged, however, that this made it possible to sell the book at an extremely low price, thus ensuring its wide distribution.

---

1 Lithophytes, psammophytes, chersophytes, eremophytes, psilophytes, sclerophytes.

2 According to Cowles, it would have been preferable to classify many conifers with the lithophytes and psammophytes and others with the oxylophytes, the mesophytes.

Warming, as numerous examples demonstrate, was adept at taking photographs and drawing illustrations. The new German edition of the 1918 *Lehrbuch*, with a foreword by Warming and co-authored by Graebner, is a voluminous work of almost 1,000 pages, with an updated bibliography and almost 400 illustrations: drawings and photographs by several authors, including Graebner and Warming, Børgesen, Raunkiær and Engler. It is also worth noting that the quality of the printed photographs had improved in just 10 years. His 1892 *Lagoa Santa* was illustrated with drawings, sketches and photographs, as were his publications on the Arctic. At the time, he had benefited from generous support.

Cowles concluded:

> *Plantesamfund* will be for all time the great ecological classic, and the English volume now before us is the most important ecological work in any language. It is at the same time an old and a new book, a translation of the masterpiece of 1895 and a compendium of ecological thinking of 1909. Warming has been contributing to ecology for more than forty years, and is the undisputed Nestor of the subject, but unlike many a Nestor, Warming incarnates the ambitions and plasticity of youth (Cowles 1909, p. 152).

The venerable and wise Warming, like the King of Pylos in the *Iliad* and *Odyssey*, offered advice. Nestor receives Telemachus, son of Ulysses and Penelope, seeking information about his parents. Cowles, evoking “the plasticity of youth”, was he thinking of the young Martin Vahl, who had come to assist the master?

Cowles was a vigilant and critical admirer of the first generation of American ecology. He recognized Warming as the master who reached out to them by publishing in English and gave value to their early work.

13.2.2. *A disappointed admirer*

Tansley’s early writings were favorable to Warming’s ecology. From the 1909 English edition onwards, things took a turn for the worse.

In a small 1904 study on “The problems of ecology” in *The New Phytologist*[3], which highlighted the terminological and conceptual difficulties of nascent ecology, he discussed the different senses in which authors understand plant ecology and agreed with Schimper. Thus:

3 Tansley founded the magazine in 1902 and was its editor-in-chief until 1932.

> Ecology is simply the topographical physiology of plants (or, as Schimper has it, (Plant-Geography on a physiological basis) – using physiology in its wider sense to include all functional relations – and this is a fact we shall do well to keep continually in mind (Tansley 1904, p. 192).

He then discussed the causes of the geographical distribution of plants, the main one being phylogenetic. He explained that plant families are confined to certain regions because their ancestors originated there. They then evolved locally. There is no reason to believe that there was anything inherent in the make-up of these plants that would prevent them from conquering other territories. In many cases, they have had to fail because of competition with other species, in their attempts to extend their areas of distribution. In some cases, they are so specialized that they are unable to change environment because they are dependent on particular environmental conditions (Tansley 1904, p. 193)[4]. Other families are cosmopolitan, for example, algae and fungi. This may be due to their age, which has given them the ability to colonize virtually the entire surface of the globe.

These two types of plant, either confined to one region or cosmopolitan, form associations. Differences in the distribution of these associations are linked to the nature of the soil (itself dependent on geology), the presence of water in the soil (dependent on its nature and the presence of water reserves in the vicinity), rainfall, humidity and air movements, soil salinity and temperature and light differences.

Nothing struck Tansley more than the strong similarity of associations in similar habitats, in different parts of the world. "The morphological and physiological characters of the plants which make up the different associations found in these different environments are, as is well known, the result of *adaptation* to the conditions of life", as demonstrated by Warming, Schimper, the Montpellier plant geographer Charles Flahault and his students, and Paul Graebner in Germany.

Today, the study of plant associations in their relations with each other and with their environment is the main subject of ecology. To study them, Tansley defined a program with several stages: describe (in terms of floristics and physical conditions), characterize and name associations, study their variations and the transitions from one to another and experiment. "It involves careful and patient observation and experiment, and the application and adaptation of the methods of ordinary physiology to the solution of these special problems" (Tansley 1904, p. 194, p. 196). This also involved complex cartographic work – begun by Flahault in France – and

4 Examples of large families that are exclusively tropical: Myrtaceae, Melastomataceae, Scitamineae; exclusively temperate: Ranunculaceae, Papaveraceae, Cruciferae, etc. Highly specialized families: Podostemaceae and Balanophoraceae.

by the creation of specialized laboratories dedicated to experimental studies in ecology.

So far, so good for Warming. Tansley did not pit him against Schimper, they complemented each other.

We may have expected a positive attitude towards the 1909 edition. But Tansley had a hard edge. By switching from Cowles' analysis to his own, there was a change of style and, above all, certain innovations and scientific options in *Oecology of Plants* were, in his view, indefensible.

Tansley admitted that the lengthy developments on growth forms in the introduction to *Oecology of Plants* were welcome and that their study led to profound morphological, anatomical and physiological investigations. He appreciated section III (Warming 1909, p. 96 *ff*) on anatomical adaptations, despite the fact that it dealt with now-familiar knowledge. He noted the importance of the contributions of Raunkiær, too little known in the United States.

While he acknowledged that Warming deftly dealt with difficult and obscure issues, he felt that this quality was at the root of some of its weaknesses. His conclusion was clear:

> The great feature of the "Œcologische Pflanzengeographie" [from 1896] was the consistent treatment of plant-communities in relation to water of the environment, as seen in their primary classification "into mesophytes, xerophytes and hydrophytes, with halophytes standing apart, though closely allied to the xerophytes" (Tansley 1909, p. 219).

But the introduction of numerous local descriptions and communities of plants in different parts of the world "has led to a considerable loss in the quality of unity of treatment and '*Uebersichtlichtkeit*', [clarity, in German in the text] which were such conspicuous features of the original work" (Tansley 1909, p. 219). He took up some of the difficulties raised by Cowles and pointed out others.

He regretted that Warming did not introduce the principle of succession into his classification, the cause of the multiplication in the number of plant formations. This would have reduced the problem also raised by Cowles. He used the example of conifers to illustrate this point. Taking into account the principle of succession would also have removed certain practical difficulties. This problem was also raised by another botanist in an October 1909 issue of the *Geographical Journal*, published by the Royal Geographical Society of Great Britain. The author signed his name "GFSE". This was probably the South African botanist of Franco-Scottish origin

George Francis Scott-Elliot, a personal friend of Patrick Geddes, one of the pioneers of ecology in Great Britain and initiator of original thinking in human ecology.

In fairness to him, Warming did point out that "between the different groups there are very gradual stages of transition", but, although he developed this point, effectively he did not take note of it in constructing his classification system (Warming 1909, p. 135).

13.2.3. *The quest for the Grail*

Tansley took the opportunity to analyze the proposal for a universal classification of plant communities by two German authors (Brockmann-Jerosch and Rubel 1912) to continue to settle scores with the obsolete physiognomic classifications, inherited from the wanderings of Grisebach, which led to "absurdities". The two German authors did not classify Schimper in the category of "old phytogeographers", because of the advances he made in plant physiology, but they did criticize him for placing his system within a general physiognomic framework which they considered outdated.

They believed that a large number of plant formations were the expression of species affinities, rather than of climate and other ecological factors, citing Warming for his excessive emphasis on these factors.

Tansley did not agree with all the proposals put forward by the two German botanists, far from it. He even found their argument unconvincing, and their system remained "ecologico-physionomic". However, their study illustrated, at the beginning of the 20th century, the tensions between different approaches to ecological classifications: "ecologico-physionomic", "topographico-physionomic" or even "physionomico-floristic" (Tansley 1913, pp. 27–28, p. 30).

The quest for the Grail, for a natural, universal ecological classification, was far from over! It was the transposition to ecology of the long history of taxonomy also in search of universality.

Tansley considered the best system to be Schimper's, although still very imperfect. Agreeing with Cowles, he regretted that the 13 classes of plant communities proposed by Warming grouped together plants that had little to do with each other or were insufficiently studied. He also pointed out that some of them only had in common the fact that they could not be included in other categories. He referred to the vague definition of class no. 13, that of mesophytes (intermediates between hydrophytes and xerophytes). They formed a "rubbish heap".

Next came the thorny issue of choosing the fundamental unit of ecology (Tansley 1909, pp. 221–222). Tansley considered that Warming rightly rejected the term "formation" in 1896, as it was used in different ways by different authors. But he did not subscribe to Warming's new taxonomy, which made formation the fundamental unit of ecology and association a subdivision of it. When Warming discussed the question of the pre-eminence of chemical or physical factors on plants, according to Tansley, he was asking the wrong question, which should be physiological.

To Warming's fundamentally physiognomic – and secondarily floristic – approach, Tansley preferred Schimper's physiological approach, or the dynamic ecology from the physiological angle of the American Frederic Edward Clements. However, he felt that the time was not yet ripe for a comprehensive work. It was first necessary to embark on the study and research of the natural units of vegetation, without leading to confusion of the great global syntheses (Tansley 1909, p. 224; Drude 1913)[5].

Tansley wrote in 1920:

> If we admit, like all those who worked on the subject admit that vegetation forms natural units which have an individuality of their own, and that these units owe their existence to the interaction of individual plants of different species with their environment, then it becomes clear that a mere study of the distribution of species as species cannot form the basis of the science of vegetation. We have instead to focus our attention on the vegetational units themselves.
>
> [...] No one can doubt that life-form represents one very important result of the response of plants to their environment [thanks to Warming!], but life-form does not run everywhere parallel with effective environment, because plants have different ways of maintaining themselves under identical conditions, and different life-forms exist side by side. Detailed analyses of species and life-forms are indispensable to our knowledge and characterisation of the units of vegetation, but neither can be made the actual basis of classification (Tansley 1920, pp. 120–121).

5 Others, such as the German plant ecologist Oscar Drude, identified the 1909 publication as the first work of general ecology.

13.3. Leaving the road

In section XVII (Warming 1909, p. 348 *ff*) on the struggle between plant communities, the last in *Oecology of Plants*, Warming asserts that plants have a special ability to adapt to new conditions in order to be equipped to live in harmony with their new environment: "Plants possess a peculiar inherent force or faculty by the exercise of which they directly adapt themselves to new conditions, that is to say, they change in such a manner as to become fitted to existence in accordance with their new surroundings" (Warming 1909, p. 370). The two and a half pages that follow develop examples that did not convince Tansley – far from it.

It could not be, for specific reasons that go back to Warming's teleological vision of natural phenomena. The reaffirmation of his positions by the author of *Oecology of Plants* exasperated Tansley, words not being too strong: "The appeal to a generalized 'inherent force or faculty' which enables a plant to react usefully to itself whatever the nature of the stimulus really does seem to amount to a frank abandonment of scientific method" (Tansley 1909, p. 225).

He went on to say that it was better to say that they did not understand the phenomenon of adaptation than to appeal to a mysterious faculty, if they did not believe that natural selection was the effective general cause of adaptation. Tansley was no fan of mystery in science. Finally, agreeing with Cowles, he found the terminology unclear for students, despite the book's excellent English translation. For him, keeping the spelling *oecology* was pedantic.

More than a decade passed between the first German translation and the English edition of his treatise on ecology. Was Warming double-crossed, or surprised, by Schimper, first with his German edition of 1898, then with the English edition of 1903?

One thing is certain: whatever time he had planned to devote – or could spare – to getting the English edition underway, he was delayed by certain difficulties. It may be recalled that the beginning of the 20th century was for him a period of investment in ambitious publications on Scandinavian and Arctic botany and botanical geography.

For the English edition, the challenge was twofold: on the one hand, to be as visible as Schimper to an English-speaking readership, particularly among the generation of North American ecology specialists, and second, to publish a revised and updated version, in a particularly fertile period for ecology. To put it another way, Warming risked being left behind, both in terms of visibility and theory, if it remained with the German edition.

Cowles remained a critical admirer of Warming. For Tansley he was the "father of modern ecology" (Tansley 1911). But it was Schimper who made ecological botanical geography accessible to English-speaking researchers, teachers and students.

But some things just did not fit!

Tansley criticized him for not taking physiology into account, which led to certain misinterpretations or badly-posed questions (because they remained in the fields of anatomy and morphology, physics and chemistry), his non-adherence to natural selection as the fundamental explanatory principle of adaptation, not to mention its mystical dimension.

13.4. Extension

In the foreword to the German edition of 1918, Warming wrote that he was taking over the organization of *Oecology of Plants*, with a few changes, including bringing the definitions of the terms formation and association in line with the conclusions of the International Botanical Congress in Brussels (1910), which he had attended. The plan of the 1918 work, co-authored by Graebner, was once again modified in relation to previous editions.

The subdivisions of the first section on ecological factors reflected a desire for clarification: edaphic factors, climatic and water factors. Warming seemed to have heard the criticisms of his 13 "classes" of formations. These were no longer to be found. Nine "series", which were highly composite to say the least, mobilized morphological, anatomical, physiognomic, climatic and edaphic criteria.

It was doubtful whether the identification of "the sub-xerophilous formation series with grassy soil" (Warming 1918, series no. 8, p. 807 *ff*), "the cold desert series" (Warming 1918, series no. 5, p. 694 *ff*), etc. would improve clarity and precision. The grouping of mesophiles and hygrophiles in the same series (Warming 1918, series no. 3, p. 525 *ff*) did not resolve the difficulties raised in 1909 by Cowles and Tansley. Finally, the substantive or nominal form halophyte appeared at the same hierarchical level (the series) as the adjective forms mesophile and hygrophile. This contributed to the confusion of Warming's taxonomy, as these terms refer to different ecological approaches.

The terms halophyte and halophile, hygrophyte and hygrophile, etc. are not synonymous. Halophytic plants are adapted to salty environments, but some can also grow on non-salty soils (e.g. species of the genus *Atriplex*, which form seashore bushes). In this case, they are facultative halophytes, which are difficult to classify as halophilous (salt-loving). Others, on the contrary, are exclusive or obligatory halophytes, which can be considered halophilous, such as glasswort.

The 1918 edition bears witness to the enormous output that had marked the previous 20 years, to the extent that a single author could not absorb it, hence Warming's heartfelt thanks to the contributors who had taken on several subjects: marine plankton and marine vegetation, steppes and grasslands, Australian deserts and illustrations.

This had been a characteristic of Warming's since his early work on Brazil. Beyond mere convention, which is a matter of the researcher's ethics, the terms employed were intended to express a sense of gratitude. Contrary to what some of Warming's contemporaries had written, the author of *Plantesamfund*, *Lehrbuch* and *Oecology of plants* himself asserted that he never claimed to publish a synthesis. Tansley rightly considered that the time had not yet come.

The first ecological synthesis, based on ecosystem theory, was published in 1953 by the Odum brothers. *Fundamentals of Ecology*, which became the standard textbook for all ecologists and students of ecology.

14

Limits and Potentials

14.1. The glass ceiling of Warming's ecology

14.1.1. *Ecology based on physiology*

Opposition and comparison, not always in Warming's favor, with the Strasbourg botanist Andreas Franz Wilhelm Schimper was based on essential theoretical points.

Schimper is cited in the 1896 *Lehrbuch* bibliography, the 1902 edition adds his imposing work of over a thousand pages: *Pflanzengeographie auf physiologischer Grundlage* (*Plant-Geography Upon a Physiological Basis*), published in 1898. He is also cited in the 1909 English version.

Schimper came from an Alsatian family of botanists. He studied natural history at the University of Strasbourg (then German) from 1874 to 1878, the city where his father Wilhelm Philippe, cousin of Karl Friedrich – already mentioned among the proponents of an "Ice Age" – held the chair of geology. Schimper planned to pursue research in mineralogy, but he also worked on the histology, ecology and geographical distribution of plants. Importantly, he "received a solid training in experimental physiology, principally in photosynthesis and in plant metabolism" (Worster 1992, p. 220) and taught at the University of Bonn, then Basel (Switzerland).

When his 1898 book was published, he was far from Europe. He was part of a major nine-month German expedition on the *Valdivia* (Canary Islands, Cameroon, South Africa, Kerguelen, Indonesian islands, Maldives, Seychelles, Red Sea), a postal freighter refitted as a vessel for oceanographic exploration in the southern seas. The expedition, financed by the Emperor, was led by a zoologist, Professor Carl Chun from the University of Leipzig.

The study of remote vegetation and his initial training steered Schimper's work towards plant physiology. He returned in 1899. However, he had contracted malaria in the Cameroon estuary and Tanzania, which claimed his life on September 9, 1901, at the age of 45.

The first part of his 1898 work dealt with the effect of ecological factors on plants (water, heat, light, air, soil, animals). The second part was a classification of the world's vegetation into regions (cold, temperate, hot), formations and smaller units. The third part involved a systematic review of this vegetation. His work also described how plants occupy new areas and become permanently established.

The three regions were identified on the basis of isotherms, which divide the vegetation into more or less parallel zones, a division already used by others. Macrothermal vegetation depends on the high temperatures that prevail below the tropics and perishes at or slightly above the freezing point; mesotherms require an alternation of high and low temperatures, with differences in their sensitivity to sub-zero temperatures; microtherms withstand very low temperatures and complete their life cycle in a short time. Schimper also grouped plants according to their similar biological forms: lianas, epiphytes, saprophytes and parasites (Schimper 1898, p. 208, p. 227, 1903, p. 192, p. 209). On first reading, the plan of his work seems to follow the same logic as Warming. On the contrary, it was not intended as a theoretical approach to ecology.

Innovation lies elsewhere.

Schimper was categorical: "The ecology of plant-distribution will succeed in opening out new paths on condition only that it leans closely on experimental physiology" (Schimper 1898, p. 4, 1903, p. 6). This condition was emphasized in 1903 in the foreword by Percy Groom, editor of the English translation at the University of Bonn, where he had met Schimper. His collaborator for this English edition was Bailey Balfour. As we have seen, they took charge of the English edition of Warming's treatise.

They were committed to Schimper's approach. Groom wrote in the foreword to Schimper's book:

> Schimper from the first insisted on the employment of methods as strict as those used in solving morphological and physiological problems. And he showed himself the master of oecological method by his critical and concurrent use of three distinct modes of investigation, namely, of observations on the comparative morphology including histology, on the physiology, and geographical distribution of plants (Schimper 1903, p. 12).

Acot notes that "for Schimper it was a matter of systematically studying the way in which the environment has a physiological impact on plant organs" (Acot 1988, p. 56).

This point, the most innovative, was addressed by the author in Chapter 1:

> It is usual to designate the plants of moist localities as hygrophytes and those of dry localities as xerophytes, but in this due attention is not paid to the fact that the characteristics of organisms are physiological, those of habitats are physical, and that there is no necessary connexion [sic] between these two groups of characteristics. In reality, a very wet substratum is quite dry to a plant if the latter cannot absorb water from it, whilst a soil, that appears to us to be quite dry may supply sufficient water to many accommodating plants. *A distinction must therefore be made between physical and physiological dryness and between physical and physiological moistness*; only the physiological characteristics need be considered in plant-life and in geographical botany. *A hyrophilous vegetation corresponds to physiological moistness and a xerophilous vegetation to physiological dryness*.
>
> Xerophytes and hygrophytes are connected by transitional forms which obscure the boundaries between them as two great oecological categories.

This division is an "arbitrary convention":

> It appears therefore necessary to place in a third category *all plants whose conditions of life are, according to the season of the year, alternatively those of hygrophytes or xerophytes*. All such plants, including, for instance, the great majority of the plants composing the Central European flora, should be termed *tropophytes*. *The structure of their perennial parts is xerophilous, and that of their parts that are present only in the wet season are hygrophilous*.
>
> The classification of plants into hygrophytes, tropophytes and xerophytes is the first step towards the physiological comprehension of the earth's vegetation and its components, the formations. Extensive districts, for instance a large portion of the tropical coasts and mountain ranges, are marked by the prevalence of hygrophytes; others, such as steppes, deserts and polar zones, of xerophytes; and others, again, for instance the greater part of the northern temperate zone, of tropophytes. There are *hygrophytic*, *xerophytic and*

> *tropophytic climates*. Every climatic district exhibits, besides the corresponding oecological type of vegetation, one of the two other types in certain localities, because the properties of certain kinds of soil weaken, or strengthen, the influence of the climate. The influence of the soil can be called *edaphic*. *There are climatic and edaphic hygrophytes, xerophytes and tropophytes* (Schimper 1898, pp. 4–5, 1903, pp. 2–3).

The qualifier *phile* designates an organism capable of living in a given environment: dry (*xero*), humid (*hygro*). The substantive form *phyte* refers to plants with these aptitudes, which Schimper implies are physiological.

This was a challenge to Warming's classification, not on technical or syntactical points, but in depth, as his criteria were considered by Schimper as non-operational. Not only were the boundaries between xerophytes and hygrophytes blurred by Schimper, but also the plants changed category – as it were – according to the seasons, the nature of the soil, variations in climatic factors and so on. What is more, some parts of the same plant may be hygrophilous and others xerophilous! Finally, knowledge of a soil's ability to retain water said nothing about a plant's capacity to obtain that water.

Schimper took up the concept of training formulated in the early 19th century, but Flahault notes that with him, formation became "a gathering of plants determined by the qualities of the soil and the conditions of the environment; it is physiological; for him [Schimper], there are climatic formations and edaphic formations" (Flahault 1900, p. 442).

Warming's system was also challenged by Schimper about the separation between xerophytes and halophytes, which follows directly from what has been said above about the distinction between physical and physiological soil dryness.

Groom experimented and observed, not by questioning the environment:

> [that] Schimper showed the xerophilous nature of the leaves of epiphytes, halophytes, and alpine plants, which dwell in physiologically dry places, whether the physiological drought be due to scanty water supply, or to unavailability of the water by reason of its salinity, or to external influences promoting transpiration. [...] Schimper pointed out the existence of many epiphytes which are not xerophytic, but may even be hygrophytic in nature, and he further correlated this with the fact that these particular plants exist as epiphytes only on very moist and shady parts of tree-trunks, and

> consequently require no careful provision against excessive transpiration (Schimper 1903, pp. 13–14).

Based on experimental evidence and solid arguments, Schimper continued to deconstruct Warming's system. Like Warming, he multiplied the examples and, skillfully, repeated his comparisons of Greenland and Saharan vegetation, but with different interpretations. He argued that Warming failed to see that "external factors are physiologically equivalent". In other words, physical drought and cold have the same effect on plants, namely, to make access to water more difficult. Plants then develop the same devices to reduce transpiration. These devices were described by Warming (Schimper 1903, pp. 679–680).

14.1.2. *A solid argument*

To measure the potential of Schimper's ecology on this question of physiology, it is useful to go further into Warming's thinking, by considering sections four and five of the 1896 *Lehrbuch* on the community of xerophytes (*Xerophytenvereine,* 130 pages) and that of halophytes (*Halophytenvereine*, 20 pages). Xerophytes (*Xerophytenvereine*, 130 pages) and halophytes (*Halophytenvereine*, 20 pages) as he characterized them.

He opens the section on xerophytes with some general remarks (Warming 1896, Chapter 1, pp. 178–180): xerophytes are "all plants that are morphologically or anatomically specially equipped to withstand more or less prolonged drought". They "live on extremely dry soil and air" or in conditions that "make water supply difficult" (cold and salt). "In addition, the importance of transpiration (evaporation) must be taken into account, [it] depends partly on external factors and partly on internal factors". In extremely dry environments, increased transpiration can be particularly harmful.

On first reading, therefore, he seemed to recognize that drought could be linked to difficulties for certain plants in obtaining water, due to their equipment. He even considered internal factors. Ultimately, however, his interpretative model did not lead him to accept Schimper's physiological explanations, as we will see.

The purpose of Warming's approach was to explore, as comprehensively as possible, the impact of water as an ecological factor on plant morphology and anatomy. This exploration is heavily referenced. It once again demonstrates his in-depth, up-to-date work at a cellular and intracellular level, and his mastery of microscopic techniques. Among his many examples, he highlighted two extreme cases: lichens, which are highly resistant to long-term drought, while others can only withstand it for short periods, and the adaptation of annual plants, with their short cycle, to alternating wet and dry seasons. He observed fine anatomical particularities. For example, the contents of the protoplasm cells (in a living cell, this is the cytoplasm and nucleus).

He identified two types of adaptation modalities of xerophytes: 1) decrease in transpiration during the critical period; 2) development of special means, partly for collecting water, partly for storing it. He pointed out that the two modalities often coexist within the same species.

The first point concerns transpiration.

To minimize transpiration, plants have several strategies: periodic reduction of surface area; movement according to light intensity ("photometric movements"); profile position (compass plants); leaves and shoots with reduced surface area; anatomical protection against strong heating and transpiration (pilosity, etc.) (Warming 1896, pp. 178–180, 1902, pp. 187–189). Compass plants belong to a group of species in the *Composaceae* family. They orient their leaves along a north–south axis, in a plane perpendicular to the ground, which protects them from the scorching sun around midday, thus limiting transpiration (wild lettuces, cup plants, which often pass for sunflowers at flowering time).

According to Warming and many of his contemporaries, aquatic plants do not transpire (Zeiger et al. 1987, p. 1, p. 50)[1]:

> The effect of their environment on aquatic plants is marked not only by the absence of transpiration, but also by other peculiar water conditions, such as absorption and diminution of light, dissolved air, movements, floatability and other characteristics (Warming 1896, p. 93, 1902, p. 97)[2].

The second point concerns water collection and storage.

The salt glands (glandular hairs) of many desert plants, including species of the tamarisk family, secrete salt, which solidifies during the day, giving them a whitish color and protecting them from transpiration. At night, the salt dissolves because it absorbs moisture from the air and gives back water, even in the absence of dew, and the plant becomes green again by absorbing this water (Warming 1896, pp. 196–197, 1902, pp. 205–206).

---

1 The submerged leaves of aquatic plants have no stomata, while floating plants have stomata on their upper surface.

2 In Chapter XXVIII "Adaptations of aquatic plants" (hydrophytes) (Warming 1896, pp. 96–100) of *Oecology of Plants*, Warming confirms, completes and updates his statements.

For xerophytes living in extreme drought, the arrival of water is an event. As it is sometimes short-lived, they need to know how to take advantage of it and mobilize all their water absorption capacities. Here too, adaptive phenomena can occur on a cellular scale, affecting the walls and contents of epidermal cells. Some plants have water reservoirs, like epiphytes and Bromeliads, such as *Tillandsia usneoides*, commonly known as angel hair or old man's beard, or desert plants. Warming cited plants from subtropical Africa (*Diplotaxis harra*, *Stachys aegyptiaca*, *Convolvulus lanatus*; a yellow crucifer, a nettle, a bindweed).

In terms of the halophyte community, Warming explained that it could be found along coasts, on the shores of inland salt lakes, near salt springs and on ancient seabeds that have not been washed away by rain. Their geographical distribution is very wide. European glasswort could be found on every continent in the temperate zone. *Glaux maritima*, a small pink-flowered salt meadow plant, is found on the northwestern coasts of Europe and the salt steppes of Tibet. In tropical zones, the mangrove is a halophyte characteristic of mangroves. These particular environments are found in the tidal zone or at the mouths of certain rivers. The stilted roots of mangroves give them a strange appearance.

Generally speaking, we are dealing with a poor flora, with well-identified families (Warming 1895, pp. 249–250, 1896, pp. 292–293, 1902, pp. 304–305)[3]. Perennial species are few in number and smaller in size. These formations have their own floristic particularities, but their physiognomies present similarities. With this type of reminder, Warming reaffirmed the physiognomic foundations of his ecology.

In the *Lehrbuch* of 1896, he relies on two earlier publications by Schimper, from 1890 and 1891, on "Means of protecting leaves against transpiration" and "Indo-Malayan coastal flora".

Schimper pointed out that the halophytes of the Indo-Malayan coastal flora often have xerophilic characteristics, even though they grow in waterlogged soils. In other words, he revealed adaptations to drought in halophytes that thrive in salt marshes. The explanation lies in the systems that limit transpiration and hence salt absorption, thus reducing the risk of reaching concentrations that are toxic to the plant.

Building on his work in plant cell physiology, he planned to supplement it with experiments on the absorption of salts by plants (sodium chloride, saltpeter and other mineral salts used in their nutrition). He made the link between "physiological

3 Chenopodiaceae, Aizoaceae, Plumbaginaceae, Portulacaceae, Tamaricaceae, Frankeniaceae, Rhizophoraceae, Zygophyllaceae, Cruciferae, Caryophyllaceae, Euphorbiaceae, Cyperaceae, Graminaceae, Malvaceae, Primulaceae, Asparagoideae, Compositae, Betulaceae, Fagaceae, Piperaceae, Urticaceae, Rosaceae, Ericaceae, Araceae, etc.

properties" and "systematics" and was already putting forward the notion of a physiologically dry environment (Schimper 1890, 1891, p. 26, p. 143).

Warming then asked the same question as Schimper, but doubted the relevance of the Strasbourg botanist's answer:

> What is the reason for this strange correspondence between plants growing in very dry soil and very dry air and these plants of which it appears that many will grow in similar conditions (continental salt steppe vegetation), others on the coast grow where the air is not at all dry and the soil may be rich in water, sometimes even flooded by the sea […], or grow steadily in water like mangrove plants? Schimper (IV, V) [he refers to the bibliography] tried to give the answer. He first points out the harmful influence that the salt present in the cellular sap exerts on assimilation and on life as a whole; the salt becomes a poison for the plant, in that it is easily absorbed in too large quantities and then has a lethal effect. To prevent too much salt from being carried away by transpiration and being excreted in the cells, the plants, according to his explanation, would be protected against excessive transpiration and therefore many protective devices would be designed against this. It must seem doubtful that this explanation is correct; because if, although extremely slow and weak but lasting, the transpiration continues, the masses of salt in the plant have surely accumulated in the plant until a specific rate for each species is reached (Warming 1895, p. 249, p. 253, 1896, p. 292, p. 297, 1902, p. 305, p. 309).

In other words, the mechanisms proposed by Schimper were unconvincing for Warming; for him, the end result was the same: the plant will die, because the lethal level of salt will eventually be reached. He made no attempt to explore this idea of the existence of "numerous protective systems", which he himself had observed.

Another point of disagreement between Warming and Schimper concerned tiny structures that could be observed under the microscope: stomata. These are the pores found in plants, generally in the epidermis of leaves, through which 90% of water is transpired. They regulate exchanges with the environment by being more or less open.

This question is therefore related to those above, but Warming transposed it to the field of ecological classification. For him, a plant unable to close its stomata could not be classified as a halophyte. This was disputed by Schimper. Warming also claimed to have observed that halophyte stomata are rarely encrypted (these crypts reduce contact with the air) and only weakly lignified. These arguments were in favor of a distinction between halophytes and xerophytes.

He therefore went to the very end of the argument to reaffirm the relevance of his positions. Whether talking about transpiration, water collection and storage or stomata, he stuck to morphological, anatomical and mechanical considerations, but not functional ones in the physiological sense of the term.

Debates on transpiration in relation to salt absorption were lively among specialists contemporary with Warming and Schimper (Muenscher 1922). By the end of the 1970s, the issue was far from resolved. Paul Binet, of the plant physiology laboratory of the University of Caen, notes that "the stomata of halophytes are still very poorly known" (Binet 1978, p. 83).

14.1.3. *A selection of references*

When it comes to the physiological foundations of ecology, we could say, trivially, that Warming missed the boat, even though he himself had considered the conditions that can make water supply difficult and was familiar with the most recent work. But things are not quite that simple. How can we explain this scientific position?

The path towards physiological explanations was opened up in the 1830s by botanist René Dutrochet on osmosis, pursued in particular by de Vries in the 1880s. Osmosis involves exchanges between two liquid solutions of different concentrations, separated by a semi-permeable membrane. Under these conditions, water always moves from the more dilute to the more concentrated medium. Consequently, if the external medium is too concentrated, the plant will not absorb the water, and may even reject it until it dries out. Each species has its own osmotic balance, its own threshold. This is where physiology comes into play. To limit water loss or even reverse the flow, cytoplasm salinity can be increased, cell membrane permeability can be regulated, and specific membrane flow mechanisms can prevent water from escaping. As we have seen, Warming himself reported on active salt concentration and even excretion mechanisms in plants of the tamarisk and mangrove family.

Unless we fall into a form of naive inductivism, which postulates that scientists observe nature without any pre-established mental or theoretical schema, we can usefully refer to the philosopher and sociologist of science Thomas Kuhn. When a scientist adheres to a theory, they defend it with all the more conviction as they belong to a community of peers who refer to it. This leads to ignoring certain observations or results, to consider them anomalies or, to save their reference model, to multiply ad hoc hypotheses, the exceptions that confirm the rule. This is the strength of the paradigm in Kuhn's sense: a representation of the world based on models shared by a community (Kuhn 1962; Chalmer 1987). In this case, this position on physical dryness, which has become dogmatic, was originally taken up by Warming's compatriot Schouw.

His own observations on anatomical and morphological adaptations common to halophytes and xerophytes could have led Warming to have doubts: succulent species, thick, fleshy, leathery, waxy leaves, mucous cells with mucilage, greater development of palisade tissue (upper parenchyma of horizontal dicotyledonous leaves that participate in nutrition), etc.

To rescue his model, he relied on experiments conducted by German botanist Julius von Sachs and French botanist Pierre Lesage. This time, he did not refer to Sachs' 1875 book, translated into English in 1890, *History of Botany*, which had a major influence on the physiological approach, nor to his handbook of experimental plant physiology (*Handbuch der Experimental-Physiologie der Pflanzen*), published in 1865, yet included in the references of the 1896 *Lehrbuch*. Warming relied on an earlier study by Sachs, dated 1859, *Über den Einfluss der chemischen und der physikalischen Beschaffenheit des Bodens auf die Transpiration* ("On the influence of the physical and chemical nature of the soil on transpiration").

It did not escape Balfour's notice in his foreword to the English edition of Schimper's *Plant Geography* (1903) that Sachs' work after 1859 (the year of publication of *On the Origin of Species*) was "illuminated by Darwin". As for the results of Lesage's experiments, falling within a Lamarckian framework, Warming was satisfied. In particular, he found that "salt acts morphologically in much the same way as sunlight", that salty environments have an effect on the morphology of a plant such as *Lepidium sativum* (garden cress), and that increasing the salt content of the soil leads to a reduction in leaf size and an increase in leaf thickness.

These observations are objective and the experimental results reproducible, but they say nothing about physiological mechanisms. Warming's choices, on the contrary, say something about his scientific stance, expressed forcefully: "what Sachs demonstrated in 1859" about the difficulty of plants to draw water from salty soils is far more probable than the explanation given by Schimper (Warming 1895, pp. 250–251, p. 253, 1896, pp. 293–294, p. 297, 1902, pp. 306–307, p. 310). With Schimper, we move from an ecological botanical geography to an ecology of plant distribution – to a functional ecology (Buchmann 2002, pp. 3–4).

He explored the physiological capacities of plant species, which he believed were the result of a Darwinian evolutionary process, while Warming focused on the study of morphological and anatomical adaptations of plants in response to environmental constraints, as defined by Lamarck. Their opposition on scientific issues was therefore underpinned by the adherence of the first to Darwin's theory and Lamarck's theory for the second.

Today, xerophytes and halophytes are still identified as distinct categories, but group together species known as "extremophiles". They can subsist in highly selective environments where water availability is particularly low. The former are dry or arid, the latter saline soils. "Although tolerant of different abiotic conditions, their response mechanisms to these constraints are similar insofar as the saline constraint translates in the plant into a situation of physiological water deficit" (Sahli 2016–2017, p. 5).

14.2. Two diverging lines

At the beginning of the 20th century, the floristic lineage was essential compared to the physionomic lineage of plant ecology, with the first great European and American schools: the phytosociological school of Zurich-Montpellier, the Uppsala school and the Chicago School. At a practical level, reference to the floristic lineage requires as exhaustive an inventory as possible of the species encountered in the field, in order to identify the characteristics of associations. This lineage was promoted by the Swiss botanist Augustin Pyramus de Candolle at the beginning of the 19th century.

We have seen that Drude was part of the physiognomic tradition of botanical geography by proposing a classification system for plant formations. In 1913, he took stock of physiognomic ecology. He situated it in reference to Humboldt and its history continued with Grisebach, with whom he had worked in his youth. He took up Warming's concept of "physiognomic life form". Some are "decisive. [They] have shaped the settlement units of land and water". The characteristics of plant formations are the result of "the ecological determination of their external conditions [and] the determination of the physiognomy unifying physiognomy of life forms".

Drude was determined to develop a phytogeographical research program for the 20th century based on several axes:

– the combined effect of climatic and edaphic factors and "physiognomic life forms in regular association";

– dynamic colonization processes and the "epharmonic adjustments" (*epharmonischen Anpassungen*) that enable new equilibriums to be achieved;

– the effects of the struggle for existence;

– consideration of phylogenetics, in the sense of the study of relationships between living beings (individuals, populations, species);

– strengthening the physiological foundations of plant ecology.

His physiognomic approach – modernized – incorporates concepts from the ecology of the United States with a floristic dimension at association level, the physiological foundations laid by Schimper (he stated that the link between morphology, physiology and the preservation of the physiognomy of life forms was proven), a Darwinian concept (already integrated by Warming in 1909) and the phylogenetic factor, downplayed by Warming (Drude 1914, pp. 10–12).

A physiognomic tradition continued and flourished in the 20th century with European authors (German, American, Austrian, Danish, French, Russian, Swiss, Ukrainian).

Swiss botanist Eduard Rübel retained the association's definition proposed in 1910 by the International Botanical Congress in Brussels, focusing on physiognomy, floristic composition, ecology (climatic, edaphic and biotic factors). These three elements formed the basis of his natural classification of associations, in line with Warming and Drude. Indeed, associations are not identified on the basis of their total floristic composition but on the basis of one or two dominant species, and sometimes using a geographical term. Rübel took Warming's idea of a classification of "association groups" which he developed in collaboration with the Swiss botanist Marie-Charlotte Brockmann-Jerosch, a specialist in alpine phytogeography and the subject of her 1903 thesis on the history and origin of Swiss alpine flora (Brockmann-Jerosch 1903; Offner 1930; Rübel 1930). They took part in the first international phytogeographical excursions (British Isles 1911, North America 1913, Switzerland 1923), initiated by Tansley, who was very active in overcoming national and linguistic barriers in plant ecology.

Botanist and geographer Krasnov of Kharkov University (Ukraine) links phytogeography to the climate classification of the Russian-German botanist Köppen, in turn based on Grisebach's classification. Köppen used plants as meteorological instruments capable of integrating variable climatic elements (Köppen 1900; Köppen and Geiger 1930). The physiognomic classification by Danish botanist Christen Raunkiær is based on adaptive forms, or "life forms", in response to the bad season: cold in temperate zones, dry in tropical zones. It is therefore based on climatic conditions, the physiognomy and the vertical stratification of plant communities. In terrestrial environments, there are several strata: endogenous (soil microfauna and microflora), bryoid (mosses), herbaceous (grasses, ferns, horsetails) and arborescent (trees, tree ferns in tropical countries, palms).

Raunkiær proposed a first version in 1904, and then refined his classification. The revised 1934 version is still in use (Raunkiær 1904, 1934; Molinier and Vignes 1971, pp. 60–77). It is based on the parts of the plant that ensure its survival. Five categories were created:

– phanerophytes (visible plants) are trees and shrubs whose dormant buds are located more than 25 cm or 50 cm above the ground depending on the region (arbitrary height corresponding to the average snow depth);

– chamaephytes (ground plants) are woody or herbaceous plants whose dormant buds, protected by snow, are less than 25 cm or 50 cm from the ground (e.g. blueberry);

– hemicryptophytes (half-hidden plants) are perennials whose buds are at ground level, protected from wind and cold by snow and dead leaves (e.g. daisies);

– geophytes (plants in the ground) are perennials with bulbs, tubers and underground rhizomes;

– therophytes (season plants) are annuals that survive the bad season as seeds.

French algae specialist Feldmann, proposed a classification of biological types in 1938. In the Mediterranean, the peak season for algae is summer (and not spring as in the case of flowering plants) and the off-season, relative, is fall (Feldmann 1938; Molinier and Vignes 1971, p. 159).

14.3. A school of tropical ecology

In the 1950s, a school of tropical ecology developed the most advanced and innovative physionomic approach, remobilizing the "life zone" concept created at the end of the 19th century.

This school was founded in Costa Rica by Renselaer Holdridge, botanist, graduate in dasonomy (the science of trees) from the University of Maine, doctor in ecology from the University of Michigan, with a specialty in dendrology (recognition and classification of woody plants). Arriving in Costa Rica in 1949, he met those who would become his students.

The context is that of the promotion of tropical studies by universities (Florida, Kansas). The scientific objective was coupled with an interest in the resources of what was not yet called biodiversity, in terms of food and health. Holdrige, along with other US scientists, was commissioned to select a suitable site for a tropical biology station. Costa Rica was chosen. Holdridge had been working since 1949 at the Inter-American Institute of Agricultural Sciences in Turrialba (on the Atlantic slope), and the University of Costa Rica was active. Finally, since that time, the small, pacifist democracy without an army has enjoyed great political and social stability.

Holdridge together with Costa Ricans and Americans founded the Tropical Scientific Center (*Centro Científico Tropical*, CCT) in San José, the capital of Costa Rica, in 1962. In 1972, the Center created the Monteverde Cloud Forest Biological Reserve in the Tilarán mountain range in northwest Costa Rica. It manages a network of reserves (Los Cusingos ornithological reserve, San Luis biological reserve, Kelady forest reserve) and develops research, environmental education, environmental services and ecotourism activities, setting up management and conservation plans for tropical forests. The CCT has conducted ecological studies in 25 countries in Latin America, Africa and Asia. Several of its members, students of Holdridge, met between 1996 and 2002 with French botanist Beauvais, including doctors Joseph Tosi and Humberto Jimenez Saa, Julio César Calvo-Alvarado, then Executive Director of the CCT, provided numerous documents and information on the "life zone system" and its genesis.

Holdridge identified important historical milestones, representative of the European physiognomic tradition. These include the authors mentioned above, notably Humboldt, Grisebach, and above all Warming and Schimper, the authors of "two books which have had a great influence on all subsequent studies of the vegetation of the world" (Holdridge 1953, p. 2). He also drew on a physiognomic tradition developed in the United States from the early 19th century. The ornithologist Merriam listed 56 authors in a synthesis published at the end of the century (Merriam 1892, 1895). They were mainly botanists until the 1850s, with zoologists dominating in the years 1850–1870. This gave the American physiognomic tradition a specific character.

Merriam's system is based on the work of zoologist Theodore Nicholas Gill, president of the Biological Society of Washington, who proposed characterizing zoological realms: nine terrestrial and five maritime. For the zoologist Agassiz, already cited for his trip to Brazil, the factors that determine the terrestrial domains are temperature and humidity. Reducing synonymies – the American physiognomic tradition has the same problems as the European one – Merriam defined seven life areas for the non-tropical part of North America, which he subdivided into life regions, also called "life zones", at the borders of which he identified transition zones, determined by local variations in ecological factors. Within a "life zone", animal and plant populations are identified. Distributed throughout the country through natural history museums, this classification became popular.

Geographer Neumann, of the University of California, shows that the theory of life zones was criticized in the first decades of the 20th century, then declined due to the difficulty of determining these zones in the field, of applying the theory on a continental scale, because of its reductionist approach linked to the sole consideration of climatic factors and the sum of temperatures to explain the geographical distribution of species (Neumann 2017, p. 37 *ff*).

Holdridge took up and redefined the concept of "life zone" by integrating other ecological factors and the concept of "climax". In a short article in 1947, he presented his "life zone" concept for the first time, and defined a vast program: "determination of the world's plant formations from simple climatic data" (Holdridge 1947). The life zone system was published by the CTT in 1967, with photographs by Tosi (Holdridge 1967).

The system defines plant formations on the basis of three climatic parameters: temperature, precipitation and humidity. The temperature factor is expressed by the biotemperature by eliminating temperatures unfavorable to vegetation growth. In the tropics, only temperatures between 0°C and 30°C are taken into account. Below and above this range, the data is eliminated. In temperate zones, the growing season is limited mainly by low temperatures. Depending on the author, the limit is around 6°C to 7°C. Biogeographer Laurent Simon uses an average of 6°C, bearing in mind that the vegetation period varies from species to species, depending on their thermal requirements (Simon 1998, 2000)[4].

Holdridge translated the humidity factor by calculating potential evapotranspiration, a concept proposed by Thornwaite in 1948. It is used to calculate the theoretical quantity of water that can be returned to the atmosphere by the vegetation cover and by the soil and its community, given a sufficient supply of water.

Finally, in Holdridge's system, "life zones" are determined by mean annual biotemperature, total annual precipitation, the ratio between mean annual potential evapotranspiration and mean annual total precipitation.

Vegetation, animal activity, climate and soil create a unified combination with a recognizable physiognomy. In the field, these physiognomies are identified by researchers, who, with practice, end up fixing mental images of these "life zones".

This is the essence of the physionomic approach conceived by Humboldt at the beginning of the 19th century. Physionomic characteristics are objectified by quantifiable data, to which Holdridge gave a graphic representation (see Figure 14.1).

---

4 For some conifers, activity picks up around 3°C to 5°C.

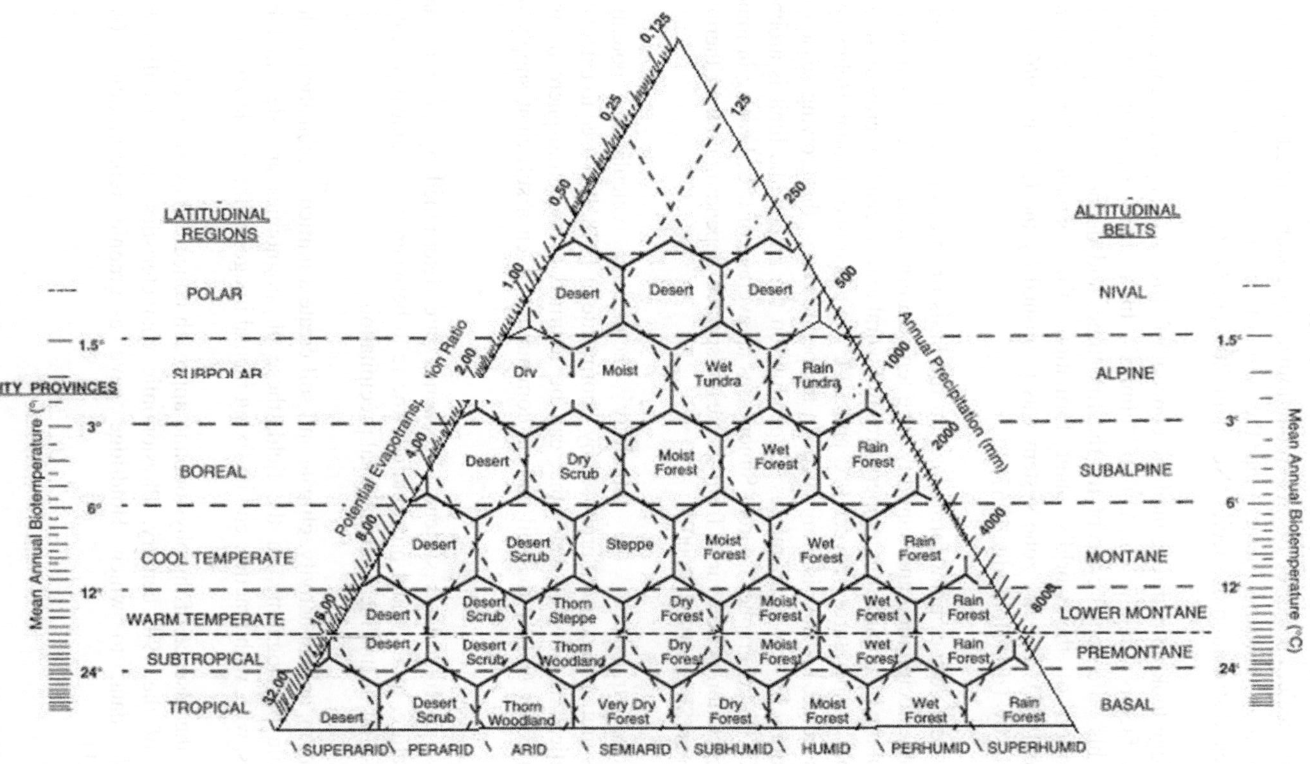

Figure 14.1. *From the diagram for the classification of life zones or plant formations of the world (Holdridge 1967, p. 17)*

The three climatic and bioclimatic parameters were used to construct this graphical representation, based on three graduated axes that draw hexagonal units: the "life zones" (Sawyer and Lindsey 1963; Holdridge 1967, p. 17; Beauvais and Matagne 1999, 2014)[5].

The life zone is defined as a natural territorial unit in which vegetation, animal activities, climate, physiography, geological formation and soil are interrelated, in a recognized and unique combination that has a typical appearance or physiognomy.

The system has found a major application in the modeling of possible evolutions of vegetation due to climate change. The studies make it possible to make recommendations on the sustainable use of the soil, on the conservation and restoration of plant formations, taking as a framework the "life zones" to be prioritized. A study carried out in Peru in 2017 covers the 16 "life zones" of the country, another from 2018 on the 13 of Guatemala (Aybar Camacho et al. 2017; Pérez Irungaray et al. 2018, p. 13).

Phytogeography, when linked solely to the physiognomic approach, could be considered pre-scientific compared to floristic approaches (phytosociology, phytoecology) (Alexandre and Génin 2011). However, in the 1950s and 1960s, the tropics, particularly from the study of the tropical forest – the most complete and complex model of an ecosystem – gave rise to ecological approaches which demonstrated the conceptual fruitfulness of the physionomic lineage, when it integrates floristic dimensions and dynamic ecology.

Warming's ecology contains all of these potentialities.

14.4. A dynamic ecology

14.4.1. *A long tradition*

The most innovative part of Warming's treatise, in reference to European geobotanical traditions, is found in the last section of his treatise, dedicated to "the struggle between plant communities". We will return to this subject in the next chapter, on the relationship between ecology and evolutionary theories.

---

5 The two-dimensional representation defines some 30 hexagons that delimit the characteristics of a "life zone". A three-dimensional reading reveals some 130 life zones for the entire globe. Hexagons are hexagonal prisms. Their distal end emerges in a latitudinal region, with the proximal end on the side of the floor chosen by the observer. This highlights "related life zones".

An analogy has sometimes been used to distinguish the ecology born in Europe from that of the New World. The former was initially interested in adapted states. In a way, it photographed different states but did not address the question of the transition from one to the other. The second, known as dynamic ecology or succession ecology, focuses precisely on these transitions. This ecology characterizes the approaches of the first American specialists. It draws on the analogy with the cinema (admittedly, these are still images in rapid succession, but the phenomenon of retinal persistence gives the illusion of continuous movement).

We can, however, identify European work that anticipates the development of succession ecology. As early as the end of the 18th century, the German botanist Karl Ludwig Willdenow, who worked on many South American plants, opened up new perspectives:

> The decay of those mosses and smaller plants produces, by degrees, a thin stratum of earth, which increases with years, and now even allows some shrubs and trees to grow in it, till finally, after a long series of years, where once barren rocks stood, large forests with their magnificent branches delight the wanderer's eye (Willdenow 1805, p. 393; Nicolson 2013, p. 96).

The historian of ecology Jean-Marc Drouin lifts the veil on a little-known European tradition which uses the concept of succession, borrowed from the alternation model in agriculture. It was developed in France in the early 19th century by Adolphe Dureau de La Malle, a landowner in the Orne region of France (Drouin 1994; Acot and Drouin 1997; Acot 1998, pp. 9–18; Trudgill 2012a).

In 1825, he noted that he observed:

> Social plants [cranberries and heathers] alternate several times and follow one another in turn. […] This alternation of the various families, genera or species of plants has been offered to my eyes a hundred times for trees, shrubs, sub-shrubs and bushes in the uprooting of hedges (Dureau de La Malle 1825, pp. 360–361).

Botanist and explorer Auguste de Saint-Hilaire – cited by Warming in *Lagoa Santa* in 1892 for his work on the primitive vegetation of Minas Gerais – and the botanist and phytogeographer from Clermont-Ferrand, Henri Lecoq, also addressed the question of vegetation succession (de Saint-Hilaire 1837; Lecoq 1854–1858; Kerner von Marilaün 1863, p. 239 *ff*).

With regard to the forests of the Alps, Warming took up one of Kerner von Marilaün's observations. The soil was first colonized by species with flying seeds, and then light-loving trees appeared before shade-seeking ones. The communities gradually spread to the most suitable locations. Further on, he gave examples of peat bogs and heaths that turn into forests if drained naturally or artificially. For other examples, Warming referred to several of his contemporaries who paved the way for the approximate dynamics of vegetation or provided data and insights that make it possible.

The focus was on certain environments, such as lakeshores, peat bogs and dune complexes. The practical applications of this research, in agriculture, are often present. Warming had been familiar with these environments for a long time, having traveled the west coast of Jutland during his teenage years as a student at Ribe.

For peat bogs, he gives three bibliographical references to Weber in the 1896 *Lehrbuch*, and eight in the 1902 edition. At the time, Weber was a botanist at the experimental peat bog station in Bremen (Lower Saxony), a center for agricultural research into the use of peat. He did pioneering work on the development of moorland and on the flora of peat bogs, in a study of botany in the service of agriculture: "On the evolution of bog vegetation under the influence of culture in relation to practical questions" (Weber 1894). Raised bogs are acidic wetlands, poor in mineral salts, so the conditions are extreme. They are fed solely by precipitation (ombrotrophy) and wind-borne minerals (Steenstrup 1842; Pokorny 1871; Klinge 1890; Stebler and Schröter 1892, p. 77 *ff*; Weber 1894; Graebner 1895; 1896, 1898, pp. 360–361, p. 364; Warming 1896, 1902, pp. 374–375, p. 378).

The 1896 *Lehrbuch* published the work of authors who paved the way for a dynamic approach to ecology (Trudgill 2012b).

Before developing his personal contributions on the question of vegetation succession and the struggle between communities, some of which were directly linked to his travels, Warming also relied on the Swiss botanist de Candolle, who wrote in 1820: "All plants are in a state of war". The quotation taken up by Warming was taken from a passage in de Candolle's work that illustrates this state of war, which is counter-intuitive for the walker, who generally has an Edenic vision of the plant world (de Candolle 1820, p. 384).

Warming wrote:

> All plants in a given country or place are in a state of war with one another. All are endowed with more or less effective means of reproduction and nutrition. The first plants to establish themselves by chance in a given locality tend, by the very fact that they occupy

> space, to exclude the other species: the largest suffocate the smallest; the most perennial replace those whose duration is shorter; the most fertile gradually take the space that could be occupied by those that multiply with greater difficulty (Warming 1896, p. 350, 1902, p. 364).

This succession approach is extensively explored in a publication on "the history of the plant world after the Ice Age". He points out that the use of peat bogs, clay layers and, to a lesser extent, archaeology enables the historical succession of plant communities to be reconstructed in Denmark by comparison with other Nordic regions, and thus parallels to be established between these regions (Warming 1904a, pp. 22–23).

In a series of lengthy passages and numerous examples, he describes the different phases in the succession of plant formations in the course of geological time, from the post-glacial period – when *tundra* reigned in Denmark – to the influence of climatic change, with, for example, the transition from moss and lichen formations to grasses and sedges, and the succession of forest formations from pine to oak, then beech, with transitions.

14.4.2. *Change and struggle*

The introductory remarks on the struggle between plant communities (*Der kampf zwischen den pflanzenvereinen*) in the seventh section of *Lehrbuch* (1896) take on their full meaning in light of developments in ecology in the 20th century:

> Up until now, we have treated plant communities as if they were static entities, in conditions of equilibrium and complete evolution, living side by side in peace with one another. However, this is what plant communities actually do: there is always and everywhere a struggle between plant communities, each trying to invade the territory of the other, and every small change in living conditions upsets the hitherto existing state of equilibrium and brings about both a disturbance and a change in the reciprocal relationships that remain. Apparently, even the smallest changes lead to major modifications in vegetation [...]. The distribution of vegetation around small lakes or ponds observed in West Jutland (Raunkiær, Warming) or the spread of Weber's "sub-formations" [peat bog study] in meadows or those of the "types or subtypes" of moors show the same thing. [...] The struggles in question have been the subject of very little investigation, so that a wide and attractive field of research is open (Warming 1896, pp. 350–351, 1902, p. 364).

These references allowed Warming to get to the heart of the matter, with chapters whose key words are "change" and "struggle". In this order, he discusses the formation of new soils (Chapter 2), soil-induced changes in vegetation (Chapter 3), changes in vegetation unrelated to soil or climate (Chapter 4), fighting weapons of species (Chapter 5), rare species (Chapter 6) and the origin of species (Chapter 7).

If a new soil is formed, plants will conquer it. If we follow the development of vegetation in all its phases, we see a series of struggles between individuals of migratory species over several decades. These situations occur particularly along coasts, estuaries and riverbeds, where materials are brought in. Glaciers also transport new materials, as do volcanoes. Fires and human intervention in general, particularly through cultivation, play an important role. Where they are abandoned, new soil is formed, with the addition of new seeds.

Figure 14.2. *A burnt campo in Lagoa Santa (August 1865) (Warming 1892, p. 7)*

Warming developed the case of free or fixed dunes, colonized by successive vegetations. In this way, the first species to thrive are replaced by others, notably by shrub vegetation. This can also be seen on the sandbanks formed by the flooding of the Danube: at first, grasses grow on the bare, damp sand, between which seeds of willow, poplar and alder germinate. Many plants with creeping rhizomes establish themselves on damp or drier sites. Then, the woody plants grow and form a forest, with trees providing shade that is unfavorable to grasses. A new soil is formed, enriched with humus. This phenomenon can also be observed in Poland's Vistula lowlands. In Iceland, river mud and sand form small islands that are then colonized. On the coasts, new soil can form, to the point where the tides no longer submerge them: the *Salicornia* zone is conquered by other plants (*Festuca, Juncus gerardi,* fescues, rushes), to form salt meadows. Earthworms cannot live in these environments. But if they are dammed up, rainwater washes the soil and the vegetation changes, making it possible for earthworms to settle in. Elsewhere than in Denmark, as in Iceland, volcanic islands are populated first by lichens, and then by mosses.

He focused on two regions close to his heart: Minas Gerais and Jutland. He returned to the case of the fires in Brazil. Forests then gave way to new cultivable land. "Fire is one of the means by which man intervenes in nature". When cultivation is abandoned, *campos* are established and, in time, a forest is reconstituted. On the poor soil of Jutland, particularly in the west, abandoned after it produced a low yield of grain, soil is formed.

True to his exhibition habits, Warming drew on numerous examples, here of plant succession, to derive a general model incorporating the notions of "change" and "struggle":

– The first vegetation is open, it always takes some time before it covers the ground.

– The number of initial species is low, but it increases. They first find a favorable location, then are forced to move as the environment closes in and becomes more "tyrannical". The environment becomes progressively more homogeneous and less species-rich.

– Annual species, initially very numerous, give way to perennials and woody plants. Annuals then occupy less dense locations.

– Species that migrate are geographically the closest. They are carried by the wind, birds.

– Light-loving trees appear, followed by shade-loving ones, never the other way around.

– The formation of distinct communities is a gradual process. "The first intermingled individuals actually belong to different natural associations". We can speak of initial, transitional and final communities. These phenomena are slow, linked to the slowness of migrations.

Chapter 4 deals with changes in vegetation while there are no changes in climate or soil. This is an interesting entry because it puts in the background the ecological factors treated in depth in the first section, for 80 pages. The origins of these changes are to be found, in his view, in migratory phenomena. Like other authors, he underestimated its potential, but was aware of the lack of studies on the subject. In his opinion, the distances covered were short: "It's difficult for birds and the wind to transport seeds over long distances". Sea currents, on the contrary, enable seeds to be transported over long distances.

The history of plants must be taken into account. He cited and drew inspiration from Alphonse de Candolle's *Géographie botanique raisonnée* (de Candolle 1855). The regions with the longest history of plant colonization are the richest. In South America, for example, *pampas* and savannas, which are more recent formations than forests, are floristically poorer. Numerous examples from around the world highlight the vast field of research that lies ahead. He cites the authors of works on the various environments mentioned (Chapter 3). He also reminds us that soil water content is essential for understanding changes in vegetation, hence the importance of transitions between hydrophilic, mesophilic and xerophilic environments.

Other factors, including animals, can also play a part in the evolution of vegetation. He also takes into account the impact of human activities, as we will see in the section devoted to his ecological project for the Kingdom of Denmark.

14.4.3. *American students*

In the 20th century, among the first generation of ecology specialists in Europe and the United States, many claimed to be a follower of Warming, whose treatise held an important place in their libraries. Among them, Cowles and Clements are major figures in American ecology. They are among the most influential of Illinois, around the University of Chicago, which also includes zoologists (Stephen Alfred Forbes, Victor Shelford, Charles Elton, Charles Chase Adams, etc.) (Acot 1988, p. 63 *ff*; Rhein 2003).

Taking up Warming's problem, Cowles wrote: "Ecologists seek to study those plant structures which are changing at the present time, and thus, to throw light on the origin of plant structures themselves". And further on:

> In the dune region of Lake Michigan the normal primitive formation is the beach; then, in order, the stationary beach dunes, the active or wandering dunes, the arrested or transitional dunes, and the passive or established dunes. The established dunes pass through several stages, culminating in a deciduous mesophytic forest, the normal climax type in the lake region (Cowles 1899; Acot 1988, p. 68)[6].

---

6 Quotations are translated into French by Acot from Cowles' thesis, p. 3 and p. 20.

Warming's remarks on the banks of ponds and puddles in western Jutland determined Cowles' choice of the environment studied for his doctoral thesis (PhD) in plant ecology: dune vegetation on the shores of Lake Michigan, one of the five great lakes of North America.

According to Clements, Warming was the first to propose an approach to the succession of dune vegetation. Other Americans would refer to Warming. Tansley thanked the master for having developed a dynamic point of view in ecology (Clements 1916, p. 23; Acot 1988, p. 63 *ff*; Trudgill 2012b, p. 854)[7]. McMillan, Chase Adams, Pound and Shelford refer to Warming's successional approach.

The concept of succession reflects the idea that so-called "pioneer" vegetation passes through various predictable stages – in the absence of disturbance – until it reaches a state of equilibrium. Stabilized vegetation then oscillates around this ultimate stage. Although it had been proposed a few years earlier, Warming did not yet use the term climax in his German edition, but spoke of "equilibrium conditions", an equilibrium that can be disturbed. "Climax" appears in the 1909 English version in connection with the tropical rainforest, which he considered to be a formation that had reached this stage (Warming 1909, p. 340).

7 Egerton pays tribute to Clements (paper presented at the annual meeting of the American Ecological Society, August 2008, Milwaukee, Wisconsin).

15

Warming's Ecology and Evolutionary Theories

In the various editions of his treatise on ecology, Warming set out his scientific views on evolution. These can be considered to be well thought-out, especially as he had already expressed his views in 1870. We can attest that "he accepted the hypothesis of evolution during the 1870s and redirected his microscopic research from the descriptive study of plant development to the examination of plant structure in relation to adaptation" (Coleman 1986, p. 185).

He returned to the subject of evolution in a popular book published in Copenhagen in 1915, in which he discusses his training:

> I can only say that we younger people who studied in the [18]50s had no idea that a doctrine of evolution had been established, because no one told us about it, and no course mentioned it. It was as if all the professors at our universities had drunk from the well of forgetfulness and had completely forgotten that such a thing had ever been discussed, or as if they were ashamed of these philosophical excesses of natural science and wanted to protect the young from such dangers (Warming 1915, p. 82).

The first part of this extract concerns his studies at Ribe Gymnasium, where it seems that ideas about the transformation of living beings over time, which had been disseminated in Europe at least since the 1830s, had no place. As we have seen, these questions were discussed with students by some of the Danish teachers Warming had studied with in the 1860s, in particular zoologists (Reinhardt, Steenstrup and Lütken). His thoughts on German universities were stranger, despite the fact that, along with England, this was the country that opened up the debate the earliest, notably with Haeckel.

The question of the origin of species, addressed by Warming to the plant world, had become a key issue. His writings give an indication of how he came to construct his answers to the "how", but also to the "why", which, according to Tansley, took him outside the realm of science.

In his 1909 treatise on ecology, he developed his ideas around three concepts: the struggle for existence, epharmony and plant plasticity.

15.1. Struggles between species and communities

The struggles between communities (Chapter 5 of *Lehrbuch*, 1896) are such that, in a way, each one works towards its own downfall, preparing the ground for the one that follows:

> The coexistence of living beings is such a complex, varied and rich reality; the many members are so closely linked that a change in one aspect can lead to far-reaching changes in others. The researcher has much to do here. Not only the varied relationships between species and ecological factors discussed in the first section (light, the species' weapons of warfare, water, heat, etc.) play a role in these changes, but also the various biological peculiarities of life forms, which cannot be said to be the direct consequence of these factors [he then refers to the multitude of factors that can come into play, whether intrinsic or extrinsic]. Sometimes one species gains almost imperceptible advantages which give it a head start over another species (Warming 1896, p. 373, 1902, p. 388).

There was also a need to take account of animals, both friends (earthworms and burrowers) and enemies (parasites, insects, rodents) of vegetation.

Migrants often emerge victorious from the struggle between communities. This is the case with spring ragwort (*Senecio vernalis*) in northern Germany. He also cited the Canada waterweed (*Elodea canadensis*), which migrated from America to fresh European waters, and the traditional Christmas tree (*Picea excelsa*) for eastern Scandinavia, which had not yet reached Denmark. Of course, these species had to adapt to existing conditions. Finally, a sort of bonus was given to the first to arrive. As a result, they are more likely to take over the territory and act as a barrier to later arrivals. *Beati possidentes* (happy those who possess), he wrote.

In this final section of the *Lehrbuch*, devoted to the struggle between plant communities, Warming partially deconstructed the certainties he established for the reader in the first section, which asserted the pre-eminence of ecological factors over plants. He explained that it is firstly because the species is intrinsically well-armed that it will have a chance of winning. Then, of course, the ecological factors must be favorable to the species, but this does not mean that the species is in an optimal situation.

The struggle between species also finds expression in floristic terms (Chapter 6). For example, some species may be rare because they are not in the right conditions or because they have just arrived – in which case their frequency will increase – or, on the contrary, because they are "relic plants" of a climatic change. This is the case with glacial relics. Only a few specimens remain, and they are becoming increasingly rare. For example, in Denmark and Northern Germany (he often refers to this territory, which used to belong to Denmark)[1]. These species are of great phytogeographical interest for determining range limits and transitions.

In the final section of the *Lehrbuch*, the Darwinian concept of struggle for existence enabled him to initiate an innovative approach to plant ecology that the successions of vegetation in space and time are an effect of the struggle for existence.

He evokes two forms of struggle: that which pits plants against their environment – we could call it a struggle against the elements, which comes under the heading of autoecology – and that which pits them against other plants. These relationships between living beings are known as synecology. The results are visible in the field. For example, he interpreted the current associations of forest trees as the result of struggles against other species. In *Oecology of Plants*, he writes: "According to Darwin there arise an infinite number of indefinite, diversified, and small variations in individuals [...] Nature must exert some selective action, and by means of the 'struggle for existence' will lead to the preservation of the most useful or fitting characters". He adds: "This explanation has recently been assailed on many sides, and does not now find so many supporters as it had when first promulgated by Darwin" (Warming 1909, p. 369).

Historian James Collins, of Georgetown University (USA), notes that in the context of the early 20th century, evolutionary theory was still under strong attack over the validity of the concept of natural selection and the role of mutations in the evolution of species (Collins 1986). In France, historian Yvette Conry has shown that the introduction of Darwin's ideas – in the sense of the transfer of a scientific theory – was not effective until World War I. Resistance to Darwin's ideas was widespread and varied in nature: scientific, ideological and religious. With regard to the concept of natural selection, which was not rejected by all scientists in France, it

1 He cites *Chamaepericlymenum* (*Cornus*), *Rubus chamaemorus*, *Polygonum viviparum*, *Saxifraga hirculus*, *Scheuchzeria palustris*, *Primula farinosa* and *Carex chordorrhiza*.

was often considered at best a secondary factor of evolution. Reducing its role in this way demonstrates the non-introduction of Darwin's theory (Conry 1974).

On the contrary, for Schimper, natural selection was the first cause. "Schimper was a strict Darwinian […] holding that the characteristics of each plant species were determined by the mechanism of natural selection" (Nicolson 2013, p. 98). His adherence to Darwin's theory can be seen in the very first lines of the foreword to his botanical geography, written in Bonn at the end of July 1898. He took into account the fact that certain selected traits may not be linked to current environmental conditions.

We have seen that, on the podium of causes of the origin of species, Warming placed natural selection in third place – as early as the Danish edition of his treatise – after adaptation and the heredity of acquired characters. He explained that selection reinforces what has already been determined by adaptation – adjustment and readjustment in the forms of plants and animals to changes in the environment – and by heredity (Coleman 1986, p. 181). While he did not reject the concept of natural selection, he saw it as no more than a kind of adjustment variable.

Sometimes he invoked it, as in *Lagoa Santa* when he pointed out that human actions (fires) and drought tend to select perennial species, sometimes he mobilized a scientist who rejected this concept, like the botanist George Henslow, who interpreted the morphological and anatomical adaptations of aquatic plants in reference to Lamarck. Finally, Warming may have simply avoided this sensitive issue in front of certain audiences, such as the Association of Scandinavian Naturalists in the 1899 lecture transcribed for the *Botanical Gazette*, so as not to offend anyone.

The strategist sometimes advances masked!

15.2. The origin of species

Chapter 7, the last chapter of the *Lehrbuch* section, is titled after the 1860 German edition of Darwin's *On the Origin of Species* (*Die Entstehung der Arten*).

By migration, Warming wrote, species adapt to or modify new circumstances. "It has also been suggested that the consequence of this modification of species may be the appearance of new species" (Warming 1896, p. 376, 1902, p. 391). He then identified what could be at the origin of the appearance of new species.

Firstly, the great plasticity of organs and parts, both internal and external. Secondly, changes in the environment not being too abrupt. Finally, useful modifications in a given environment require reinforcement. This is the case, firstly, when the external circumstances that brought them about persist, secondly when they are inherited, and thirdly through natural selection. The first reinforcement mechanism is Lamarckian, the third is Darwinian. The second raised the then unresolved question of the heredity of acquired characters and their transmission.

According to Lamarck, author in 1809 of *Philosophie zoologique*, the first general theory of the transformation of living beings under the action of "circumstances", or environment, useful modifications are reinforced by use. The law of use and non-use means that an organ or structure becomes stronger when it is used, and atrophies and eventually disappears if it is not used at all. Thus, characters acquired and strengthened over the course of a lifetime confer better adaptation to the environment. In this sense, Lamarck's theory is finalist.

Warming believed – somewhat hastily – that while the principle of the heredity of acquired characters had been debated, it was now accepted, even by a Darwinian like Julien Vesque. In fact, Darwin did not formally reject the principle of heredity of acquired characters.

The distinction between germ cells (the origin of reproductive cells) and summative cells (the body's cells, excluding germ cells) had only just been established by the German physician August Weismann, who investigated the mechanisms by which acquired traits are transmitted. But when the first two German editions of the *Lehrbuch* were published, Weismann's experiments were, by his own admission, still inconclusive, and he struggled to convince his contemporaries (Weismann 1893). We now know that germ cells are the only cells that transmit genetic heritage between generations.

We are talking here about the ways in which genome elements are transmitted vertically (between generations) and not horizontally. This has been described since the second half of the 20th century, notably between bacteria, between bacteria and fungi, or even between plants and insects.

Darwin died before the first founding works of genetics, which remobilized the pea hybridization experiments of Austro-Hungarian monk and botanist Gregor Mendel, carried out in the 1850s and 1860s (Mendel 1865).

Finally, for Warming, two questions arose as soon as changes were made in the structures and functions of an organism.

– Do external forces (physical and chemical) bring about these changes, and what are the biological consequences?

– Are these changes good or bad for life? "The second question is often answered more or less hypothetically or teleologically". Warming adopted the hypothesis that plants have the ability to adapt directly to environmental conditions. "He therefore assumes that an otherwise unknown link exists between external causes and the interest in change (self-regulation or direct adaptation)" (Warming 1896, pp. 377–378, 1902, pp. 394–395).

The teleological doctrine to which he alluded had its roots in the philosophy of Aristotle, for whom phenomena were explained by the intervention of a final cause (*telos*). This is illustrated by Lamarck's famous example of the giraffe's neck.

Why do giraffes have long necks? Fixists answer that the question makes no sense, because giraffes were created that way for all eternity. Lamarck explained that when grass becomes scarce on the savannah, giraffes have to stretch their necks to reach the foliage of the trees (final cause):

> It is interesting to observe the result of habit in the peculiar shape and size of the giraffe (*Camelo-pardalis*): this animal, the largest of the mammals, is known to live in the interior of Africa in places where the soil is nearly always arid and barren, so that it is obliged to browse on the leaves of trees and to make constant effort to reach them. From this habit long maintained in all its race, it has resulted that the animal's fore-legs have become longer than its hind legs, and that its neck is lengthened to such a degree that the giraffe, without standing up on its hind legs, attains a height of six meters. (Lamarck 1809, pp. 256–257, 1963, p. 122).

Characters acquired throughout life are passed on to subsequent generations, in accordance with the principle of heredity. Living beings thus tend towards a progressive complexification that makes them ever better adapted.

Warming adopted the Lamarckian hypothesis, teleological, of direct adaptation to environmental conditions, interpreted as a final cause. For example, he explained that changes in light induce modifications in the distribution of chlorophyll, in the position of leaf blades, in the morphology and anatomical structure of certain organs and in the life cycle of species. Annuals and biennials can become perennials. He gave numerous examples of changes within a species under the influence of different ecological factors.

For Darwin, in a population of giraffes, there is an intrinsic variability which means that some individuals have slightly longer necks than others. This modification gives them an advantage: they can continue to feed after the leaves on lower branches have been consumed. Giraffes with this advantage live longer and have more offspring to whom they pass on this trait. This variability can affect parts of the body that seem unimportant, such as the giraffe's tail: "can we believe that natural selection could produce, on the one hand, organs of trifling importance, such as the tail of a giraffe, which serves as a fly-flapper" (Darwin 1859, p. 171). However, Darwin continued that resistance to insect attacks could have an impact on the geographical distribution of certain animals.

This proved to be consistent with mutationist theory. Genetic mutations occur by chance, for reasons unrelated to the goal of better adaptation. They are causality without finalities.

As for Warming, he mobilized three types of factors to explain the formation of species:

> Direct adaptation is undoubtedly a species-forming factor of the utmost importance, but naturally not the only one. Another is what Vesque calls "phyletic variability", an inherited variability that depends on the ascendance of the species but does not depend on the environment. A third factor is Darwin's natural selection according to his hypothesis about the spontaneous variation of individuals in undefined and unplanned directions (Warming 1896, p. 381, 1902, p. 398, 1909, pp. 369–370).

Vesque considered that in non-cultivated plants, "the anatomical characters of the species are epharmonics, that is to say, characters of adaptation to the physical environment". But they mask or distort the phyletic characters, which should be taken into account by botanists when describing species. He therefore drew their attention to the risk of creating varieties or races that were merely the result of non-constant adaptive phenomena. "Distinguishing between species and variety then becomes a matter of sentiment". They lose sight of the species "in the philosophical sense", as the "ultimate phyletic form". "The ultimate phyletic form is the division comprising all plants that differ from one another only in epharmonic characters".

This is an allusion to the Lyon botanist Alexis Jordan, who, within a fixist framework, gave the tiniest observed variations the status of characteristics, leading Jordanians to create a multitude of "small species". He was at the origin of the so-called Jordanist school (Bange 2004). In France, Vesque was well known to amateur naturalists. At a meeting held on January 14, 1893, members of the *Société d'émulation du Doubs* discussed the distinction to be made between phyletic and epharmonic characters, since they do not have the same origin. The former, unmodified by the environment, reflect real affinities between taxa. The latter are the result of epharmosis, a process by which plants reach a state adapted to a given environment. Consequently, systematist botanists needed to evaluate the status of characters observed on as many samples as possible before coming to any conclusions (Parmentier 1893, p. 156*ff*).

It was important not to overlook the role of heredity and natural selection, added Vesque. "Adaptation to the inert environment is the effect of the influence of the environment transmitted by heredity and fixed by natural selection" (Vesque 1882, p. 10, p. 44, 1883, p. 111, p. 114, 1889, pp. 46–47, p. 49).

Warming took a different view. He quoted Lamarck in the very last lines of the *Lehrbuch*: "Lamarck in this respect had a sharper eye for the truth than seemingly most contemporary researchers. Adaptation is certainly one of the most powerful factors of development in the organic world" (Warming 1896, p. 381, 1902, p. 399).

What was at stake here between Vesque and Warming was not adaptation versus natural selection, but the role of these phenomena in the evolutionary process. Adaptation was the result of the former and the driving force of the latter.

Finally, Vesque was a Darwinian that Warming could live with. In a way, non-Darwinians could make use of his concept of epharmony. It was all a question of interpreting the processes that enabled reaching an adapted (adaptation) or Darwinian (selection) state. On the contrary, Warming avoided mentioning Schimper, a Darwinian with whom he probably found it difficult to negotiate, scientifically speaking.

15.3. Epharmony and life forms

Warming begins the final chapter of *Oecology of Plants*, on "The origin of species" with the theme of epharmony by making an observation based on all that precedes in his book: "The structure and whole course of development of species stand in perfect harmony (epharmony) with the environment […] species are adapted to the surrounding conditions".

He pays tribute to Darwin with "epharmony prevailing in every niche and cranny of Nature". He then posed the question: "How has this epharmony arisen? […] Old-fashioned teleologists would answer: 'Through the direct creative action and wisdom of God.' Modern research, relying upon the facts of evolution, whose foundation was laid by Darwin, gives other explanations". Recently, slightly different explanations have been proposed concerning the origin of species by Korschinsky and de Vries. For the former, it was due to sudden changes appearing in a "heterogeneous" manner, and for the latter, by "mutations". Some of these characteristics are useful, others indifferent or harmful. Epharmony intervenes by exercising its selective action. It is therefore possible that many indifferent characters (morphological, systematic) remain (Warming 1909, p. 369).

The theory of mutations is, as we have seen, associated with de Vries (Amsterdam), and also to Correns (Leipzig) and von Tchermak (Vienna). Warming cited only the former and mentions Korschinsky, a Russian botanist and director of the St. Petersburg Botanical Garden, who compiled an abundance of horticultural literature showing the importance of what he called "heterogenesis" in the production of new varieties (Korschinsky 1901; McDougal 1902; Blaringhem 1905, p. 110). This term refers to the appearance of different characteristics in individuals belonging to successive generations.

As with mutations, this raises the question of the production of new forms – coexisting with the repetition of forms similar to themselves – which appear without transition, independently of the environment and the effect of use and non-use. This phenomenon thus breaks free from the laws laid down by Lamarck.

This final chapter of *Oecology of Plants* repeats a demonstration published in 1908 in a study on "life forms" by Warming.

His strategy was particularly cunning.

Not only did he not deny Darwin's contributions, but he also paid tribute to the author of *On the Origin of Species* – as he did every time he quoted the English naturalist – whose concept of natural selection renewed discussions on evolution. According to "true Darwinism", the characters selected over countless generations are those that are useful in the struggle for existence and have "selective value". Thus, "teleology, the doctrine of opportunity, which had been almost forgotten, has suddenly regained its honor and value in a somewhat modified form".

He reminded us once again that the concept of selection was merely an assumption that variations – which are the starting point of new species – are directionless and indeterminate. Mutations have the same characteristics. He quoted de Vries, forgetting to point out that he accorded natural selection a secondary role in evolution. As a result, he considered that "mutation theory and Darwin's theory of selection are quite close to each other and in clear opposition to Lamarckism". They result in forms that appear "by chance", without cause, irregular: they can be useful, harmful or indifferent to the species. "But Nature then chooses from among them [individuals] those who are best adapted to the world, who are strongest in the struggle for existence". The result is the adaptation of living organisms to their environment, which is admired by zoologists and botanists alike, he concluded.

A magnificent attempt to restore Darwin's theory through the doctrine of final causes, with the convocation of a Nature that chooses!

He went on to thank Darwin for stimulating new work on life forms that the Austrian Darwinian botanist Kerner von Marilaün was able to take advantage of. He praised Marilaün's comments on the correspondence between forms and environmental characteristics. Warming recalled the work that, in his opinion, demonstrated the existence of harmony between structure and function. In particular, he called on the French (Flahault, Bonnier, Vesque, Lesage, etc.), botanists from Germany, England and, above all, North America (Clements, Cowles, Pound) and, of course, Raunkiær's classification of biological types.

Epharmony of life forms manifests itself in morphology, anatomy and the whole periodicity of life's phenomena. In the great plant formations, we observe an "epharmonic convergence". In heather moorland, for example, the "ericoid life form" dominates, recognizable in different species with heather-like leaves. He cites *Erica*, *Calluna* and *Empetrum*. Convergence is due to environmental (climatic, edaphic) and historical factors.

"Understanding the harmony of life forms is a step towards understanding the origin of species" (Warming 1908, p. 7, p. 27, p. 29, pp. 31–33).

The die had been cast!

15.4. Plant plasticity

The theme of plant plasticity was developed at length in his work. In *Lagoa Santa* he writes:

> I respond briefly and incidentally to the question that arises, namely whether these adaptations to the environment should be considered as a guarantee, resulting from natural selection, against evaporation, or whether they owe their origin to the modifying action of forms, exerted directly by the environmental conditions. I adopt the latter view. Still, some of these morphological particularities such as leaf dwarfism, spinosity, etc. can arise from a direct influence. It will have been so, no doubt, also with the vegetation of the *campos* in which thousands and thousands of generations have succeeded in fixing the characteristics of adaptation directly acquired (Warming 1892b, p. 313).

This answer – brief and incidental – concludes a descriptive paragraph on "the xerophilous nature *of campos* vegetation". For him, direct adaptation to the environment and heredity of acquired characters explain the morphological peculiarities observed. As for the plasticity of the inner and outer parts of plants, he asserts in *Plantesamfund* (1895) that it can be at the origin of new species.

In his "famous" *Philosophie zoologique,* Lamarck assumed, before Darwin, that species undergo variation. According to Lamarck, changes occur continuously. The great diversity of forms is due to two factors: time and the nature of the environment. The organism must adapt when confronted with different local conditions. The same situation occurs when it migrates. These phenomena were studied by Étienne Geoffroy Saint-Hilaire in animals, Warming pointed out, but Lamarck assumed that the direct action of the environment on living organisms was produced above all in plants.

Geoffroy Saint-Hilaire began referring to Lamarck in the 1820s to formulate his own theory of the transformation of living beings over geological time (Corsi 1988, p. 231). He opposed the fixist Cuvier. Balzac dramatized their controversy, siding with Geoffroy Saint-Hilaire in *La Comédie humaine.*

Warming did, however, level one criticism at Lamarck – at a time when the question of the heredity of acquired characters was the subject of heated debate: "Lamarck assumed without discussion that the new forms arising in this way transmit the acquired traits to their offspring and are thus preserved by generation; but this is the weak point of the hypothesis, and it is today a question which is much discussed and is still open".

But when it came to the adaptation of plants to environmental conditions, his mind was made up, and in 1909, he spoke of himself in the third person, a convention that made him one of the leading scientists on the subject:

> Warming assumes that plants possess a peculiar inherent force or faculty by the exercise of which they *directly adapt themselves* to new conditions, that is to say, they change in such a manner as to become fitted for existence in accordance with their new surroundings [...]. He thus assumes that between external influences and the utility of variation there is a definite connexion, which is of obscure nature (*self-regulation* or *direct adaptation, epharmosis*).
>
> It is certainly a fact that the plant is extremely plastic, and that external factors can evoke numerous changes in it. This is proved by

> numbers of facts and experiments recorded during the last few decades by Costantin, Volkens, Lothélier, Stahl, Vöchting, Schenck, Lesage, G. Karsten, Frank, Dufour, Vesque, Bonnier, Askenasy, Klebs, Massart, Göbel, Lewakoffsky, Graebner, and others, who have investigated the morphological and anatomical plasticity of individuals. The result of these investigations has been to show that by change in external conditions there is set up a course of development *tending to adapt the plant to its environment in precisely the same manner as plants or plant-communities growing under natural conditions normal to them are adapted* (Warming 1909, p. 370).

Thus defending the principle of direct adaptation, which he accepted in the name of a connection "of obscure nature" between the external environment and the usefulness of the modified character, interpreted as a finality. He drew on the work and experiments of numerous scientists, then used examples to reinforce his positions.

The authors cited in this extract are German and French botanists, phytogeographers and physiologists from Germany and France (with the exception of one Belgian) who explored or sought to demonstrate experimentally the validity of the hypothesis of plant plasticity and the heredity of modifications. Following in the footsteps of zoology, botany entered the laboratory, where the aim was to guarantee the scientific validity of the results obtained.

15.5. Famous and controversial experiments

Bonnier (nine bibliographical references in *Oecology of Plants*), member of the *Académie des Sciences*, professor of botany at the Sorbonne and author of floristic books still in use, developed an experimental program within a neo-Lamarckian framework.

Neo-Lamarckism is mainly based on two interrelated notions: the plasticity of living beings and the recording through heredity of changes acquired throughout life, which are then passed on to descendants. These individually acquired traits add up to become those of the species. Neo-Lamarckism established a functional link between organism and environment. The experimental route, which Lamarck was unable to take, was to provide a scientific explanation for these phenomena, by linking morphology and physiology. Neo-Lamarckism was also promoted by biologists in the USA and Russia (then the USSR) (Lenay 1997; Loison 2010, 2012).

In 1878, Bonnier and Flahault spent several months in Sweden and Norway on assignment from the French Ministry of Education (16 references in *Oecology of Plants*) and established comparisons between Scandinavian and Alpine plants (Bonnier and Flahault 1878). Their hypotheses were to be tested by Bonnier. His most famous and controversial experiments concerned "the processes of immediate plant adaptation to climatic conditions" (Benest 1990). Bonnier set up a rigorous experimental protocol. He wrote in a popular journal:

> Since 1887, I have been growing experimental crops at an altitude of 2,400 meters on the Mont-Blanc mountain range. A single perennial plant growing on the plains around Paris is divided by its underground parts into various tufts. These clumps are separated into two batches that are absolutely comparable; the first batch is planted on a specific soil, which is the soil of the alpine station transported in a bag to the plain; the second is planted at an altitude of 2,400 meters, on soil of the same nature (Bonnier 1913, p. 31).

These experimental crops were monitored for several years. As a result:

> Samples planted at an altitude of 2,400 meters have gradually changed in form, structure and even function over the years, and most have quite rapidly acquired all the characteristics of alpine plants [...]. In another series of experiments, I was able to produce, so to speak, artificial alpine plants – while maintaining them in the plain – at my laboratory in Fontainebleau. [...] These plants, treated in this way, soon took on almost the appearance and structure of alpine plants (Bonnier 1913, pp. 32–33).

These adaptations, in just a few years were, for Bonnier, a fine demonstration of the great plasticity of plants.

Thus, the mountain form of helianthemum vulgaris gave a prostrate form (*Helianthemum vulgare*, now *Helianthemum nummularium*). On the plains, it had much longer, spreading stems and larger leaves.

Bonnier was also experimenting with plant adaptation to the Mediterranean climate, at La Garde, near Toulon. He transplanted 43 species from Fontainebleau. He observed, for example, that *Fraxinus excelsior* (European ash) acquires, in just one or two generations, characteristics of the species common around the Mediterranean *Fraxinus parvifolia* (synonymous with *F. angustifolia*, with narrow leaves).

Figure 15.1. Helianthemum vulgare. *Plain (P) and mountain (M) (Bonnier 1895a, p. 398)*

He developed an experimental botanical geography at the Fontainebleau plant biology research station, an annex of the Paris Faculty of Science, where he headed the laboratory. Pierre Lesage, also cited by Warming, a professor at the Faculty of Science in Rennes, experimented on the influence of climate on the modifications of maritime plant leaves. Julien Costantin (10 references in *Oecology of Plants*), Bonnier's brother-in-law, also followed the experimental route at the *Muséum d'histoire naturelle* in Paris. Lothelier (three references) tested changes in branches and leaves as a function of light and humidity. Belgian travel botanist Jean Massart (eight references) worked on environment-related changes in plants observed at different latitudes.

Warming also mentioned Germans: Volkens (five references) who studied plant responses to conditions in the Egyptian-Arabian desert; physiologist von Sachs tested the influence of light on photosynthesis and the production of amidon. Stahl (development of lichens and fungi), von Göbel (adaptation of forms to functions),

Vöchting (physiology and pathology tales), Karsten (11 references, Antarctic phytoplankton), Askenasy (environmental influences on plant growth), Klebs (heredity of changes under environmental influence, mangrove study), Lewakoffsky (influence of water), Gräbner (or Graebner, six references, development of plant formations in relation to the environment), Frank (plant physiology applied to agronomy), Schenck (11 references, production of aerenchyma, aeriferous tissue forming a channel in certain tropical mangrove plants), all work within a Lamarckian theoretical framework.

This large cohort was obviously mobilized on purpose by Warming, who cited Bonnier's work until the end of his life. In a study on Caryophyllaceae (e.g. carnations or silenas) published in 1920, we find references to experiments conducted 30 years earlier by Bonnier on plant adaptation in relation to climate, and the comparison of Arctic species with the same Pyrenean and Alpine species (Warming 1920).

The protocol used by the experimenters at the Fontainebleau station was considered incomplete. Indeed, we would have expected them to return the plants to their original environment, in order to verify the stability or otherwise of the new traits acquired. They would then have found that, after several generations, the original conditions would eventually erase the modified characters, in favor of those the species had initially.

In his defense, while Bonnier, by way of the railroads, was able to use a fast, modern means of transporting plants, he also had to travel over rough roads in a cart, and then climb mountains. The species he transplanted had to be hidden from the eyes of walkers so as not to be damaged. Before opening his laboratory on May 15, 1890, conditions were difficult. But these difficulties were ironed out once the biology station was opened. It was ideally located on a 2.5 ha site, near a forest and 300 m from a railway station. It could accommodate 24 workers.

Microscopic, physiological and plant chemistry studies were carried out there, and a greenhouse divided into hot and temperate, was an extension of the main building. For cultivation under glass, Bonnier, in collaboration with mycologist Dufour (two references), monitored temperature, humidity, ventilation and electric lighting for several months and recorded all values, which they did with precision (Bonnier 1890, 1895a, 1895b; Jumelle 1890; Aymonin and Keraudren-Aymonin 1990).

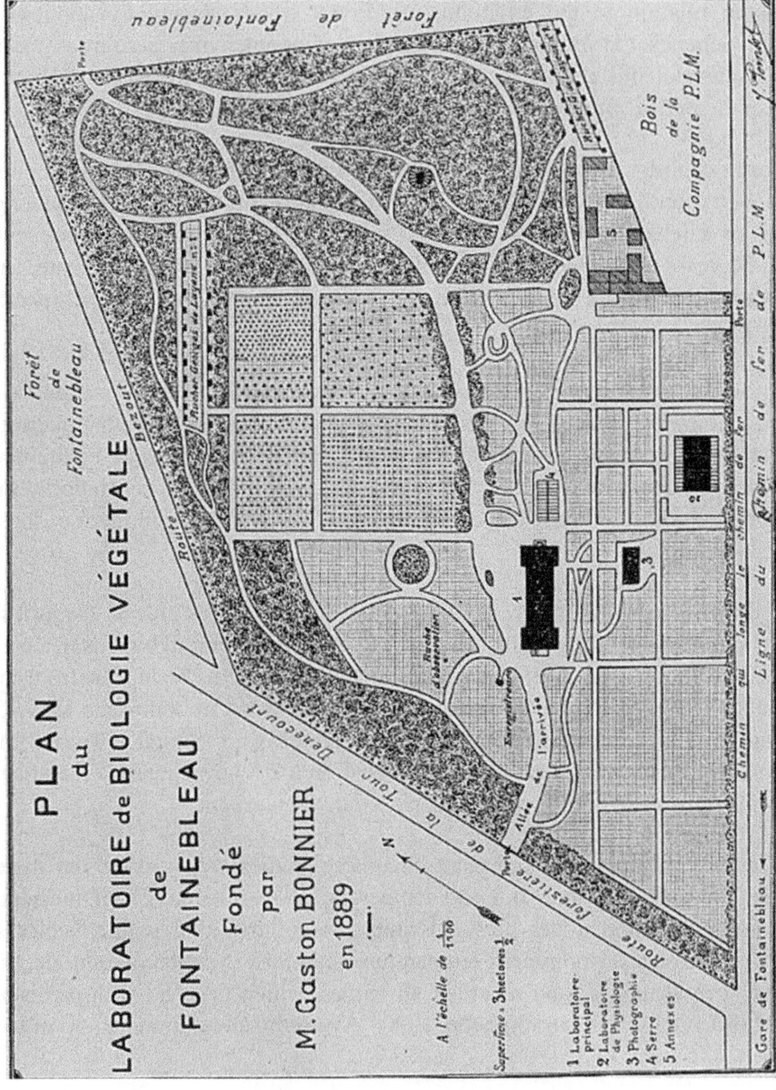

Figure 15.2. *Plan of the Fontainebleau laboratory grounds (Dufour 1914, Plate 2)*

In reality, Bonnier, like other neo-Lamarckians, did not intend to work directly on the hypothesis of the heredity of acquired characters, but on the plasticity of plants, on their adaptive responses to changes in the environment. He never said otherwise: "The result of these experiments is that the characteristics of Alpine vegetation are indeed adaptive and it can be inferred, by a very rational hypothesis, that species quite special to the high altitudes have thus gradually formed under the influence of the environment" (Bonnier 1913, p. 33). According to the Fontainebleau experimenters, it went without saying that when plants remain in the same conditions and reproduce from seeds, these modified characteristics can be found in their offspring, providing sufficient proof of the hereditary transmission of characteristics acquired during existence.

We now know that this type of response by living beings to environmental factors is not a matter of genetics, but of epigenetics. New traits are not genetically acquired, but are produced by a modification in the expression of the genome. This modification can be transmitted over several generations, but is reversible because the molecular structure of the genome remains unaffected.

Warming ends *Oecology of Plants* with an ode to direct adaptation, in line with the work of Bonnier:

> Direct adaptation is beyond doubt one of the most potent evolutionary factors in the organic world, and appears to play the leading role in the adaptation of growth forms and formations [...], from day to day new experiments go to prove that the direct action of external factors can modify the plant's nature and can evoke hereditary distinctions in form (Warming 1909, p. 373).

Even though there will always be an element of mystery, he concludes.

Before the concept of the unity and diversity of the living world was highlighted in the second half of the 20th century by evolutionary phylogenetics and molecular biology, on a theoretical level, the very notions of plasticity and character fixation remained, if not mysterious, at least contradictory.

15.6. The "Lamarckian cradle of ecology" (Acot 1997)

We may wonder why the ecologists of the late 19th and early 20th centuries sometimes accepted certain evolutionary concepts and sometimes did not. The answer is complex, linked to the heterogeneity of the players in nascent ecology, reflecting the changing and diffuse identity of ecology.

During this period, those involved in plant ecology were less likely to be evolutionists than taxonomists. The former were able to dispense with a historical approach to the phenomena they studied, while still conducting ecological studies recognized by their peers. The latter, with the rise of natural classification methods, necessarily asked questions about genealogy and the filiations between taxa. *Genera plantarum secundum ordines naturales disposita* (Genres of plants classified according to the natural method), by de Jussieu (1789), is the foundation of this approach. Ecological botanical geography is more focused on immediate environmental causes than on historical explanations, even though the latter can shed light on certain current phytogeographical situations (Hagen 1983, 1986; Collins 1986, p. 170). While specialists in plant ecology, like Warming, adopted certain Darwinian concepts, they did place their research within the framework of his evolutionary theory. Acknowledging that a theory is admirable does not automatically lead to its integration into scientific practice.

Acot has shown that some of the great names in British and American ecology of the early 20th century were Lamarckians (Acot 1997). More generally, historians of ecology are divided as to the relationship between nascent ecology and Darwinism. Their analyses are linked in particular to their interpretation of Haeckel's epistemological position in the emergence of scientific ecology and the content of *On the Origin of Species*, in particular biogeography.

In response to those who present Darwinism as an ecological theory of evolution, Acot questions the relevance of including "Darwinian ecology" in the historical current that gave rise to plant ecology (Acot 1983, p. 33).

The concept of adaptation was still the leading concept in ecology at the time of Warming. The concept of selection may have been invoked secondarily, almost by default, when certain adaptive phenomena resisted Lamarckian explanation. Finally, the study of the adapted state makes it possible to unite Lamarckians and fixists around the same ecological research program. The latter consider that this state exists from all eternity.

15.7. Didactize: translate or betray!

In a lavishly illustrated textbook, *Theory of Descent* (*Nedstamningslæren*), published for the Committee for Popular Education in 1915, Warming didacticized evolutionary theories for a readership of teachers and their pupils. This document, which complemented the analysis of the chapter in *Oecology of Plants* devoted to the origin of species, went beyond the analysis of Warming's purely scientific positions on evolution. He had been speaking on the subject to a variety of audiences for several years. In 1906, for example, he gave a series of lectures for secondary school teachers at the Frederiksborg public secondary school – co-educational since 1903 (Prytz 1984, p. 140). Retired and once again settled in his family home, the publication of his popular work was, by his own admission, the most important event of the year, the culmination of several decades of reflection and conferences given to various audiences.

It begins with a history of life on Earth, from creationist ideas and the belief in spontaneous generation to the Pasteurian revolution, via the first microscopic observations of simple living things and the birth of experimental biology in the 18th century. He constructs a narrative in the positivist tradition outlining the chronology of scientists and the continuous progress of science without forgetting, "by the way", to praise the Danes, notably Emil Christian Hansen and his work on brewer's yeast for the Carlsberg Foundation, of which Eugenius himself was an executive for over 30 years. Hansen's experiments on ferments seem to show that living organisms can change over generations under the influence of high temperature. For Warming, this was further experimental proof of the Lamarckian mechanism for the transmission of acquired traits.

The theories of Lamarck (Warming 1915, p. 51 *ff*) and Darwin (*ibid.*, p. 68 *ff*) are methodically didacticized. The former is presented as the man who laid the foundations of evolution and whose intuitions were taken up by Darwin. Lamarck had a chronological and theoretical advantage. He formulated "for the first time a clear and complete exposition of the doctrine according to which all living and extinct creatures form connected chains, whose links are linked by the bonds of kinship, the most complex descending from the simplest".

Two forces – the first transforms living beings from the simple to the complex, the second is adaptation to the environment – and two major principles, continuity, use and non-use, were set out by Warming. "Life is an uninterrupted flow broken under a constant transformation of form [...], 'species' vary (change)". He recalled Lamarck's reflections about the arbitrary boundaries created between species, even though they were linked by intermediate forms. He attributed to Lamarck an idea of the German philosopher Leibniz: *natura non facit saltus* (nature does not make jumps).

Use and non-use have an "immediate effect on living beings". They are translated by progressive changes over a long period. "All living things are subject to the laws of change". Warming's work was based on numerous examples in the human species, in the animal world – with the case of the inescapable giraffe – and in the plant world.

Lamarck was attacked and mocked, he recalled. Warming contributed, like other Lamarckians of the 19th century, to the construction of a narrative according to which the French naturalist died in poverty and rejected by all, silenced by Cuvier, except by Geoffroy Saint-Hilaire, he added. On the contrary, "when Darwin died in 1882, he was the famous naturalist whose name was on everyone's lips – a complete contrast to Lamarck's fate". The myth of the cursed scientist works well, especially in a popular work. It calls for that of the scientist ahead of his time, compared to Cuvier in particular. According to Warming, Lamarck sowed seeds that germinated half a century later with Darwin; Lyell's principles of geology having facilitated the germination of these seeds, from 1830 onwards.

Of Darwin's early voyage on the *Beagle*, he wrote: "This voyage was to determine his whole future life". He was not the only one to have made this comment, but we cannot help drawing parallels with his own voyage to Brazil, at the same age as Darwin, which also laid the foundations for a long-term scientific project published 30 years later.

He then showed that Darwin asked himself the same questions as Lamarck on the definition of species and that he too worked on domesticated animals and plants. This is where the rhetorical strategist came in: "Darwin therefore assumed three driving forces to set evolution in motion: 1) the capacity of living beings to change; 2) the heredity of new characters; 3) the selection of the most useful through a series of generations".

A little further on, he took up and developed what he considered to be the basis of Darwin's theory, in slightly different terms: "1) the mutability of species; 2) overproduction; 3) inheritance. Overproduction leads to 4) 'the struggle for existence', and then, as a probably necessary consequence; 5) selection on the part of nature". Warming thus adds the question of "overproduction". This refers to the fact that, through reproduction, living beings have a number of offspring that exceeds the capacity of the environment to meet their needs (an idea taken from Malthus). It also incorporates the explanation of mutability and the struggle for existence.

In the two preceding extracts, just a few pages apart, the presentation strategically puts natural selection in last place. This is not surprising. It downplayed natural selection, as we have seen, at least since the 1870s.

He simplified matters by not specifying that useless or neutral characters were also selected, as he noted in 1909 in *Oecology of Plants*. This omission enabled him to give a non-specialist readership a teleological vision of evolution: natural selection retains only what is favorable, by virtue of an immanent law. Warming's talents as a popularizer were undeniable. He could have given a less partial – or one-sided – explanation without complicating his discourse by presenting all the elements needed to understand a non-finalist vision of natural selection, in the spirit of Darwin.

A key to reading Darwin's work, given by the author at the outset, works evidently throughout his book: he wanted to bring out in Darwin what he took over from Lamarck. Moreover, he identified Darwinism solely with the concept of natural selection (Warming 1915, p. 4, p. 6, pp. 52–54, p. 57, p. 63, p. 69, p. 77, p. 83, p. 88). In addition, he stated that Wallace had the same theory. Darwin had the advantage over Lamarck of being able to rely on numerous facts and half a century of progress in zoology, botany, paleontology, geology, comparative anatomy and cell biology.

This was a fact, even a triviality in view of the enormous scientific progress recognized by all in the 19th century, but – contrary to what Warming alleged – it did not detract from the innovations of Darwin's theory. It only paid tribute to a true naturalist committed to the facts, a tireless worker. Although he never came to Denmark, Darwin's naturalistic work had been known and appreciated by Danish scientists since the 1830s. He even borrowed barnacles from the Museum of Zoology in Copenhagen for his work on fixed cirripedes (marine crustaceans). The Copenhagen library contains several editions of his account of his voyage on the *Beagle* and the first German translation in 1844, as well as several copies of his work on coral reefs (Kjærgaard et al. 2006).

Warming was quick to add that the English naturalist's writing was tedious and repetitive. If he wanted to dissuade Danish teachers and pupils from reading Darwin in the text to make up their own minds, even through Jens Peter Jacobsen's translations, he would not do otherwise (Jacobsen and Møller 1893)!

It seems that he was not a follower of Rousseau's pedagogical methods, who wanted to give his pupils the means to form their own opinions. As a result, he skewed the discussion by immediately directing the answers:

> Darwin does not set out to explain the cause of changes (although he also assumes that they arise from changes in the world), but he assumes that living things change from unknown causes, and then that nature comes along and selects from these random varieties those best suited to the altered species [...]. It's an old question whether ducks got their webbed feet because... they went into the water (Lamarck), or whether they went into the water because they had webbed feet (Darwin) [...]. Lamarck's explanation is apparently more profound than Darwin's, because it must lead directly to the important scientific question of how environments can act to change living things (Warming 1915, pp. 84–86).

Darwin's Copernican revolution, which placed living beings at the heart of the evolutionary process, had not reached Warming, who persisted in giving the environment a central role.

He added that Darwin changed his mind in the 6th edition of *On the Origin of Species*: "Darwin even took up Lamarck's idea that 'habit', 'use' and 'non-use' are quite important as driving forces in evolution".

On this point – deliberately highlighted by Warming – Darwin, in the successive editions of *On the Origin of Species*, after having crossed swords with his contradictors for several years, came to admit – or at least he found it impossible to reject – this Lamarckian principle. Finally, Warming organized his argument around "indirect evidence of descent". This process enabled him to provide up-to-date knowledge in a variety of fields, the meaning of which was given by the Lamarckian framework he had laid down earlier.

In the part on "comparative theory of forms" (comparative anatomy), he deals with the question of organs or analogous parts (e.g. insect and bird wings) and counterparts (bird wings and those of bats) in distinct taxa. These notions are due to Geoffroy Saint-Hilaire. He mentions the vestigial or, on the contrary, strongly developed organs (interpreted in a Lamarckian framework by the law of use and non-use) and the idea of the unity of the compositional plan of vertebrates: they "descend from the same basic form, and are transformations of this". This phenomenon is also observed in plants. He takes the example of the leaf and the peel of onions, which result from morphological and functional modifications, thus alluding to his former research on organogenetics.

His "history of the development of the individual" mobilizes the field of embryology (ontogenesis), marked by the work of the Darwinian Haeckel, whom he cited, and for whom ontogenesis recapitulated phylogenesis. Warming saw this as an illustration of the Lamarckian logic of the appearance of simple living forms, then embryogenesis, like phylogenesis, gave rise to a progressive complexification (Warming 1915, pp. 84–86, pp. 88–89, p. 92, p. 98, p. 104).

"The epitaphs of deceased organisms" – the field of paleontology and geology – can inform us about their genealogy, consistent with the history of the Earth written by Lyell. Warming gave special treatment to human paleontology, formulating the many questions that remained unanswered. The great antiquity of man, who had to cohabit with extinct species, was recognized. "All these white, yellow, red, black and brown men. People today […] look so much alike that they are only like races of one and the same species" (Warming 1915, p. 125).

This reflection was inspired by Cuvier's monogenist doctrine: "Although the human species appears to be unique, [...] there are certain hereditary conformations which constitute what we call *races*". Cuvier distinguished between white or *Caucasian*, yellow or *Mongolian*, black or *Ethiopic* races (Bronwen 2009). Polygenism makes it possible to believe in several different humanities, the offspring of distinct original couples. It is thought that this belief can be objectified and rationalized through numerous observations and measurements, notably of the cranium in the context of physical anthropology, resulting in a taxonomy.

In the 19th century, the notion of race became widespread among naturalists and historians. Warming was a European of his time. However, unlike others, he did not introduce the idea of a hierarchy between races.

His 1915 manual provoked numerous reactions. We will just mention Johannsen, chemist at the Carlsberg brewery laboratory (1881–1887), who collaborated on several of Warming's publications, with whom he quarreled. He became one of his great Danish opponents (Glick 2014, p. 154, p. 164; Poulsen 2016, pp. 154–155). He coined the terms gene, phenotype, genotype and biotype. Although he did not quote him in an article published in 1911, it is clear that he classified Warming among the zoologists and botanists who had not emancipated themselves from the old conception of transmission. He ironized Warming's undue use of metaphors drawn from archaeology and religion. A practitioner of experimental laboratory science, Johannsen rejected metaphysical and speculative approaches to biology. He criticized Lamarck's theory – without placing natural selection at the heart of evolution – claiming that environmental factors that affect the phenotype cannot be transmitted to the genotype. He defended the idea that the genotype is built up in stages, discontinuous (discontinuity between genes), and the theory of mutation "natura facit saltus", he wrote (Johannsen 1911).

Nedstamningslæren was a great success and, paradoxically, helped to introduce Darwin's ideas on evolution, while re-energizing Lamarck's work (Shaffer 2014, p. 151, p. 158; Poulsen 2016, p. 48, pp. 154–155). Ultimately, Warming may have inspired reflective and curious readers, who would know how to distance themselves from passages where he took a stand, to read Lamarck and Darwin in the text, in order to form their own opinions.

15.8. Political and religious dimensions

When it comes to adherence to the theories of Lamarck or Darwin, political and religious dimensions can be decisive. In Denmark, Grundtvig's criticism of text-based Lutheran theology and his preference for the living word legitimized an interpretation of Scripture that created particular conditions in the face of Darwin (Hjermitslev 2011, p. 279).

The question was of interest to a cultured public, thanks in particular to the writer, botanist and poet Jens Peter Jacobsen, a major disseminator of Darwinism in Denmark – although not a Darwinist himself. He was the translator of the 5th edition of *On the Origin of Species*, published in 1871–1872 in nine small pamphlets with a print run of 1,500 copies (Darwin 1872). He wrote articles on Darwinism, influenced by Haeckel, for the radical magazine *Nyt Dansk Maanedsskrift* (New Danish Monthly), which sparked off a controversy with Ditlev Gerhardt Monrad, bishop of the Danish Baltic islands of Lolland and Falster.

Literary critic Brandes, who welcomed Jacobsen's popularization of Darwin's work, exploited his polemic and anticlerical potential. He put Darwinism at the service of his fight against the Church and conservative order. The debate, which until then had been confined to academic circles, was to ignite a politicized public, which inveighed against each other in the press. The "Brandes circle" brought together liberal and socialist students.

It is amusing to note that in his 1915 manual, Warming also tackles this sensitive subject with an anecdote to arouse interest. Cautious, he does not mention the controversy between Jacobsen and Monsignor Monrad, but the much more famous quarrel between Samuel Wilberforce, bishop of Oxford and the English zoologist Huxley, known as "Darwin's bulldog". The clergyman was deeply indignant at the idea that life could be subject to pure chance and that man would have an animal ancestry, with our ancestors resembling apes (Warming 1915, pp. 85–86). Warming did not take a direct stand. He merely denounced Haeckel's excesses on the subject and called for a form of appeasement.

The intended readership for his 1915 textbook was that of the popular higher education schools. The first *folkehœjskole* was founded in Rødding in 1844 and by the beginning of the 20th century there were more than 80. They were aimed at young men (winter sessions lasting five or six months, when work in the fields stopped) and young women (summer sessions lasting three months) aged 18–25. Tuition was paid by the week, but there was a system of scholarships for the most deserving, and the state had been subsidizing the schools since 1851. However, they remained independent and free.

The "popular high school" (*folkehøjskoler*) created by Grundtvig were aimed more at education than instruction. No diploma was awarded upon graduation. Scandinavian history, the Danish language and literature were given pride of place, and a strong sense of patriotism – exacerbated during the Duchy Wars – bathed in a Christian spirit was cultivated. Teaching, methodical and progressive, was mainly oral. The living word, in the Grundtvigian spirit, was favored. Teachers were often members of religious orders (Paul-Dubois 1909; Rothmaler 1921; Jolivet 1960, pp. 16–17; Jensen 1985). Naturalist education was given the same place as mathematics, physics, chemistry and geography.

In the 1860s–1900s, the periodicals of the Grundtvigian group, supported by the Liberal Party, took up the theme of Darwinism. Their schools of higher education tended to distinguish between faith and science. In the last decades of the century, the liberal neo-Grundtvigian movement accepted the idea of evolution. By the time of Darwin's death in 1882, satirical depictions of Darwin himself and his theory were commonplace in the Danish press. His life and work, portraits, obituaries, special issues and even short stories and novels inspired by Darwinism were widely circulated.

By the 1900s, discussions of Darwinism were part of the curriculum in secondary and higher education. In 1904, Warming gave a lecture on the doctrine of evolution at the University of Copenhagen. He expressed his sympathy to students who defended the evangelical faith against Darwinism. These were obviously not the same as those of the "Brandes circle". He did not himself become involved in militant movements, but he actively supported them, notably by delivering several critical readings of Darwin's doctrine in 1909, published under the auspices of the Students' House, founded in 1892. Warming was a member of the board. Evangelical students then joined the professor, who defended and disseminated his ideas, legitimized by his institutional position (Glick et al. 2014, p. 138, p. 150). At the time, he was one of Denmark's most prominent scientists, holding key positions both at the university and in learned societies, periodicals and the powerful Carlsberg Foundation.

At the beginning of the 20th century, he was a little quick to estimate that:

> Spirits are calmer now; many have understood that the Bible is not a natural history textbook, but a means of awakening the thought of God's greatness and omnipotence; [...] and many recognize that we can assume that, in the living world, the creation of the Earth is a process that takes place within the framework of human history; an evolution is underway, which is also as governed by fixed laws as the inanimate world of multiple appearances (Warming 1915, p. 87).

Warming believed that life in general and plant communities in particular had an economy regulated by both divine and terrestrial laws: a kind of mixed economy.

Ecological science defined the terrestrial – horizontal – laws of this economy, expressed by the nature of the relationships between living beings and their environment, but transcended by divine – vertical – laws, with providentialism in the background (Ferngren 2013, pp. 345–346). Thus, in an incomplete secularized perspective on science, for Warming the Creator remained the guarantor of *oeconomia*.

As a Christian, he had difficulty accepting that natural selection was a decisive evolutionary mechanism. He left plenty of room for theological and teleological interpretations of the evolutionary process, which he expressed in his widely circulated popular publications. Nature has intentions that will always remain obscure and mysterious. He was so convinced of this that he even concluded *Oecology of Plants* with it. Tansley, as we have seen, was shocked to find such assertions in a scientific treatise.

Conclusion to Part 4

Botanist Lauritz Kolderup Rosenvinge, Warming's university colleague, placed his "golden age" in the 1890s. Then, especially from the 1909 publication onwards, Warming did not always master the many major problems he tackled. Finally, in his view, he remained stuck in his search for explanations of plant adaptations in Lamarckian terms (Kolderup Rosenvinge 1923–1924). But on this point, he was far from alone, especially among the pioneers of ecology.

Science and religion

In Denmark, the rejection of natural selection because of its potentially atheistic implications reinforced, on the part of liberal theologians and many naturalists, their adherence to the Lamarckian concept of inheritance and to a teleological vision which included the idea of progress, coherent with the Creator's plan.

A Christian naturalist, Warming believed that science had not disproved the Bible. While he accepted the idea of the evolution of the living world and sought to understand the terrestrial laws of the geographical distribution of plants on the Earth's surface, he believed in the immanent laws governing the universe. "He has a neo-Lamarckian, morphological and religiously informed understanding of plant geography, […] a religiously informed understanding of both the human and the plant community" (Anker 2011, pp. 325–327).

Whatever the hypothesis, he wrote in *Nedstamningslæren* in 1915, it only postpones or sidesteps the big question: how did life first appear? Who are we to seek to unravel the mystery of the Almighty, creator of matter, energy, time and space?

His interpretation of Lamarck's theory offered an acceptable compromise between science and faith. It reflected a restrictive vision of evolution, "a theistic evolution" (Hjermitslev 2011, p. 294), which introduced teleological metaphysics. Warming's separation of science and religion was not fully achieved, especially in his textbooks for teachers and his talks for students.

For other Lamarckians, notably from the Russian school, the direct action of the environment as the cause of variations determined by the environment had no metaphysical dimension. Thus, Peter Kropotkin, in his scientific articles of 1910–1915, reinstated Lamarck's *Philosophie zoologique* on the terrain of materialism. There was no question of invoking mysterious forces, as Warming did. If the terrestrial causes persisted, the adaptive responses – morphological and functional – became cumulative and correlative and not accidentally useful as in Darwin's work. They thus led to progressive evolution (Kropotkin 1910; Rubinstein and Confino 1992; Garcia 2015).

The contributions of Warmingian ecology

At the end of the 19th century, when *Plantesamfund* and then *Lehrbuch* were published, the theoretical difficulties were manifold for ecological botanical geography. In particular, they were linked to the terminological and conceptual inflation that marked the laborious birth of ecology as a scientific discipline.

Warming produced a compilation of the major works and trends in botanical geography of the 19th century, clarifying nomenclature and proposing a classification that made it possible to identify natural ecological units. Dissemination of his program in Europe and the United States was achieved through the German translation of his treatise, followed by a revised and updated version in English that incorporated the contributions of new authors, more and more numerous to embrace the scientific field of ecology.

Warming broke new ground on a number of points, which ultimately helped to establish ecology as a scientific discipline.

He made a clear distinction – both conceptually and methodologically – between floristic and ecological approaches. This led Warming to place the study of plant communities at the heart of plant ecology. This concept of community was forged over some 30 years, from his first expedition to Lagoa Santa, through to his mission to Greenland, to the publication of the *Plantesamfund* in 1895.

He defended the scientific character of the physiognomic approach to vegetation in ecology: "the physiognomy of vegetation is not only of aesthetic, but also of scientific significance", he wrote in 1909 (Warming 1909, p. 137). A step back from the dominant floristic approach from the beginning of the 20th century, the physiognomic approach of ecology giving rise to fruitful programs right up to the present day.

He traced the programmatic lines of a dynamic successional ecology, explored by the first specialists in the ecology of plant succession then biotics of Chicago. While scientific ecology has its roots in botany, zoology, limnology and oceanography, botany – and botanical geography – were the first providers of concepts and methods, until World War I (McIntosh 1985, p. 30). However, Warming did not benefit from the study methodologies developed by the Americans nor from ecological research programs structured by universities and supported by specialized journals.

He created the conditions for transferring concepts from plant ecology to animal ecology, founded by British zoologist Charles Elton in the 1920s. Although attempts were made at the end of the 19th century, the field of animal geography was opened up later than botanical geography (Ebach 2022, p. 18). In France, naturalists proposed applying the methods of botanical geography to zoology; "Zoologists need only follow the path opened up, or rather trace a parallel path", wrote Charles Bruyant in the *Revue scientifique du Bourbonnais*. By analogy, "flora" and "plant cover" became "fauna" and "animal population", and "plant formation" found its equivalent in "animal formation". They tried to put these principles into practice (Bruyant 1898, p. 5, 1899). Biogeography would soon be referred to as dealing with the geographical distribution of animals and plants.

The English edition of 1909, the publication of which was delayed, built on what came before, but modified the structure and some of the content. It was updated to take account of new issues, particularly those addressed by the first generation of plant ecology specialists in Great Britain and the United States. This imposing work by the respected Professor Warming, which was widely distributed, was greeted with mixed reviews, particularly by those who had been delighted to discover the first German edition. His main European competitor, Andreas Franz Wilhelm Schimper, founded ecology on physiological foundations, which Warming did not really explore, although he did promote his own teaching of the physiological and experimental approach to botany. He remains fundamentally attached to an ecology whose foundations are physiognomic and based on the study of the adaptations of organisms to their environment, in a neo-Lamarckian theoretical framework.

What if Warming had published an English translation in 1897, in line with the original, following on from the Danish edition of 1895 and the German edition of 1896?

This is based on a regret expressed several times by Warming at having to delay the release of the English-language edition.

If his project had turned out as he had hoped, Schimper would not have overshadowed him, in Goodland's words. On the contrary, Schimper's work on physiological bases published by Schimper in 1898 would have been the mark of the dynamism of a young, innovative discipline, producing and importing concepts and methods, in this case those of physiology.

This would have spared Warming the onerous task of a belated publication, which followed a difficult twofold objective: to translate the original version into English and to make changes – a revision and updating – made indispensable by the contributions of the many new authors of the 20th century.

The 1896 *Lehrbuch* was a revelation to Tansley when he discovered it in 1898:

> The German translation was widely read in England and America and played an important part in stimulating fieldwork in both countries. It certainly did in my own case: I well remember working through it with enthusiasm in 1898 and going out into the field to see how far one could match the plant communities Warming had described for Denmark in the English countryside; and I also made the book the basis of a course of University Extension lectures at Toynbee Hall [London] in 1899.

His disappointment was all the greater:

> It was a pity that Warming's brilliant textbook was not immediately translated into English: an English edition was delayed until 1909, and the new book, under the title of *Oecology of Plants*, entirely rewritten and incorporating much new matter, not all of it very sound, was far from satisfactory (Tansley 1947, p. 130).

Freed from the problem of translation into English, assuming publication at the end of the 19th century, Warming could have devoted himself to analyzing the productions of the new generation of European and American ecology specialists and to developing certain points, on the assumption that they would all have known his treatise on ecology having read it in Danish, German or English.

Warming was over 50 when the Danish version of his treatise was published and 67 when the English edition was published. Traveling to the Americas was still hard work, even though the average journey time had dropped from 40 days in the early 1830s (23 days to return to Europe, due to westerly winds) to 15 days by 1900. What is more, steamship navigation was safer and conditions much better for passengers. The railroads made it easier for him to travel around Europe, answering invitations, taking part in conferences and organizing excursions for his students.

Now based in Copenhagen, where he held prestigious positions, he consolidated his position within major institutions, taking part in all the major international events in his field of research and helping to implement the Kingdom of Denmark's scientific and ecological project.

Figure C4.1. *Portrait of Eugenius Warming circa 1900 and his signature (Marius Christensen, public domain via Wikimedia Commons)*

PART 5

The Ubiquitous Professor Warming

16

The Institutionalization of Ecology

Warming is strongly represented and involved, essential even, in the works and debates of the period described by Pascal Acot as "institutional". This period "saw the birth of all the major schools of plant sociology, in the course of confrontations and trial and error lasting more than thirty years" (Acot 1988, p. 99).

Although he was not present at the 1st International Botanical Congress in Paris in 1900, his name was at the heart of the discussions, as we will see. His treatise on vegetal ecology was authoritative. In 1905, he attended the 2nd Congress in Vienna. On June 16, 1905, he posed in a photograph with 75 other participants in the deliberations on nomenclature. He also attended the 9th International Geography Congress in Geneva, from July 27 to August 6, 1908. He was part of Section VIII "Biological geography (botanical geography, zoogeography)" but did not include any contributions. In fact, only seven papers were presented in this section, which was very few in comparison with the other sections. On the contrary, he contributed actively to the important work of the 3rd Botanical Congress in 1910 in Brussels. The fourth of this name was not held until 1926, for the first time outside a European capital, in Ithaca, New York, two years after his death.

Specialized ecology journals were born. The British Vegetation Committee (1904) became the British Ecological Society in 1913. Its *Journal of Ecology* gave pride of place to ecological studies, which until then had coexisted with reports of excursions contributing to botanical cartography. The first issue of the *New Phytologist*, founded by Tansley, appeared in the UK in 1902. In the United States, The Ecological Society of America was founded in 1916. *Plant World*, which had taken over from *The Asa Gray Bulletin*, became *Ecology* in 1920, the voice of The Ecological Society of America.

Figure 16.1. *International Botanical Congress, Vienna, 1905. Warming is no. 41, circled in the photo (Meise Botanical Garden, Belgium, see: http://creativecommons.org)*

At the beginning of the 20th century, France and Germany did not create any new press organs specializing in ecology, existing journals included ecological publications. In France, this was the case with the *Revue générale de botanique*, edited by Bonnier, *Le Monde des plantes*, the *Académie internationale de géographie botanique*, the *Bulletin de la Société botanique de France* and numerous periodical publications by academic societies. In Germany, the Free Association for Plant Geography and Systematic Botany published works on ecological botanical geography and systematics.

16.1. "Babelic confusion"

The opening of this institutional period was marked by the 7th International Geography Congress in Berlin from September 28 to October 4, 1899, at which plant geographers launched a consultation with the aim, they hoped in the short term, of reaching agreement on nomenclature. Among the subjects discussed (mathematics, physics, industrial geography, etc.) was that of "biological geography". "The 7th Congress chose from among biogeographers living in and around Berlin, a preparatory commission whose aim would be to prepare a simple, unified system of nomenclature for plant formations for submission to the next congress", to be held in Paris in 1900 (Septième Congrès 1899, p. 30).

German phytogeographers Warburg, Engler and Drude took the initiative. They approached Flahault. Prevented from taking part in the Berlin congress, he joined them and presented a landmark "phyto-geographical nomenclature project" in Paris which was to become a milestone. The great plant geographer from Montpellier – where he founded the *Institut de Botanique* – reminded us of the fundamentals. "Phytogeographic nomenclature must apply [...] to geographical and topographical units [and] to biological units". To do this, it must first solve several problems: the synonymy of these units must be international, the rules adopted for phytogeographical cartography must be common, and the terminology must be identical. He tried to impose a cartographic basis shared by all phytogeographers, with a unified language. Flahault recommended following the proposals of several authors (Grisebach, Drude, Warming, Schimper, etc.). He proposed defining two types of unit, geographical and biological, and laid the foundations for a nomenclature inspired in particular by Warming.

The first-order geographical and topographical units were the "hot, temperate and cold zones". These were broken down into major climate-related "vegetation regions" (Northern Eurasian forest region, Mediterranean region, Alpine region, etc.), subdivided into domains, sectors, districts, subdistricts and stations. These represented "the combination of climatic and geographical factors with edaphic and biological factors, i.e. the relationship of each species with the soil and with the species with which it is associated". Differences between stations were therefore linked to fine variations in one or more factors. Their names were specific to each country, making the task of the nomenclator particularly arduous. These mutually subordinate units were complex: geographics, physiognomics (linked to climate, topography, soil) and floristics.

For biological units, Flahault tackled the question of their nomenclature by starting with "elementary units, those that populate the stations", to successfully approach more encompassing units: groups of associations, ecological series of association groups, vegetation types. The latter were determined by "the same set of climatic conditions combining in the same way". The vegetation association was the basic unit. It "is the ultimate expression of vital competition and adaptation to the environment in the grouping of species". Some are "dominant", others are "secondary" or "subordinate". This approach to association was in line with the views of several authors, including Warming, who clarified the notion of biological form (*Lebensform*, *Vegetationsform*). Flahault saw it as the equivalent of the station.

He noted the great interest of one of Warming's innovations: the "group of associations (*Vereinsklasse*) to encompass in a single whole several associations subject to the same general environmental conditions". He declared: "Phytogeographers need only follow the excellent principles laid down by Mr. Warming".

But strong dissensions remained, leading to polemics from which the floristic lineage emerged strengthened. At that time, it was "a babelic confusion, an inextricable maze" (Flahault 1900, p. 429, p. 431, p. 436, pp. 443–444, p. 438).

16.2. Lengthy preparations

Naturalists are particularly attentive and vigilant on the question of nomenclature. If a general consultation is not launched, any decision, whatever it may be, will never be followed by action. It has to win the support of as many people as possible. In 1900, as at the Congress of Vienna in 1905, this was a lost cause. Flahault and Swiss botanist Schröter were tasked with gathering the propositions for submission to the Brussels Congress in 1910.

Flahault's call for contributions was relayed by major international journals, notably *Botanische Jahrbücher* (Botanical Yearbooks), published in Leipzig. Warburg made "suggestions for the introduction of uniform nomenclature in plant geography". The following year, Engler took stock of "the progress made in plant geography" since 1899. He highlighted work on botanical geography in the hot, temperate and cold zones of the world, including those of Warming. Most of the authors cited were European. There were, however, a few references to Americans: Cowles, Clements and Pound (Warburg 1901; Engler 1902). The aim was to mobilize the contributions of all those who mattered in the fields of phytogeography and plant ecology.

At the same time as the Brussels Congress, the *Freie Vereinigung für Pflanzengeographie und Systematische Botanik* (Free Association of Plant Geography and Systematic Botany) met for the eighth time. In France, after the Revolution, the term "*libre*" (free) was used to distinguish the new learned societies from the academies of the *Ancien Régime*[1], which were subject to royal authorization by letters patent. In Germany, the associative movement developed in the second half of the 19th century according to liberal conceptions. Associations were independent of the State, and operating rules were relaxed by the law of April 19, 1908 (Scheuner 2016, p. 23, p. 25).

The association exerted its scientific and programmatic freedom. It was a constituted group in which proposals found in Brussels in 1910 were tested and discussed. It had been publishing since 1903 (*Bericht über die Zusammenkunft der Freien Vereinigung für Pflanzengeographie und Systematische Botanik* (Report on the meeting of the Free Association of Plant Geography and Systematic Botany)). Drude and Engler were on the Board of Directors. The association could be

1 The political and social system of France from the Late Middle Ages until 1789.

perceived as a think-tank and, thanks to the presence of certain members, as a form of lobbying. It held two important meetings: the first on May 14, 1910, the second on May 25. The conjunction of the calendar was not ambivalent.

On the morning of May 14, at 8:30 a.m., the session opened in the amphitheater of the Botanical Institute in Münster, made available to the audience by Correns, one of the founders of mutationist theory. The Brussels Congress opened on the same day. In fact, the official opening took place on May 15 at 2:30 p.m. in the Botanical Gardens. An extraordinary session was followed by a tour of the garden, greenhouses and herbarium, where a glass of champagne was served. Only Board members were invited to attend in the morning, from 10:30 to 11:30. Work commenced in earnest on May 16.

Members of the Free Association taking part in the Brussels Congress were able to free themselves on the afternoon of May 14 or the morning of May 15 after their session in Münster, to travel to Dortmund or Osnabrück station, some 50 km away, and arrive in Brussels for the start of the proceedings. They were accustomed to using horse-drawn carriages and trains for their travels and excursions. Their meetings were usually preceded by botanical excursions, scholarly walks and convivial meetings in fine hotel restaurants, with the gentlemen accompanied by their ladies.

The second meeting of the Free Association took place on May 25 in Dahlem-Berlin. The Brussels Congress had ended on May 22, so the delegates had three days in which to meet in Berlin, in an Oxford-style science district. They then met in Bromberger and Danzig (1911), Freiburg (1912), Leipzig and Berlin (1913).

Two points can be made about the activities of the Free Association in the run-up to the 1910 Brussels Congress. Firstly, it worked on the question of nomenclature, crossing the fields of phytogeography and systematics, and secondly, it incorporated the impact of human activities into its work.

The first question had been worked on for a long time. Drude, who chaired the 7th meeting in Hesse on August 9, 1909 (on the Rhine, about 20 km from Mainz) at the phytopathology research station, was already pleased with the progress of work on the "vegetation of the Earth" and on the "natural families of plants". Joel Hagen (University of Radford, Virginia), historian of ecology and plant taxonomy, noted that at the turn of the 19th and 20th centuries, botanists defined a new field of research which turned its back on the old tradition of natural history. Ecological botanical geography now focused on communities rather than species, on immediate environmental causes rather than historical explanations, on physiological experiments rather than morphological descriptions (Hagen 1983, 1986). The influence of Warming and Schimper is directly readable in the definition of these fields of research.

The second question was an innovative one. The contents of the newsletters consulted for the period 1909–1914 confirm that the theme of the impact of human activities was indeed being worked on. Article 2 of the new statutes submitted to the members of the Free Association at their May 14, 1910 meeting in Münster stated that the association was dedicated to the protection and preservation of the natural environment. This issue was generally ignored by European phytogeographers before the First World War (Massart 1912)[2].

Against a backdrop of rapid industrialization in Germany – in comparison with France – the second half of the 19th century saw the emergence of ideas relating to *Raubwirtschaft* (destructive economy). These were driven by geographers (Ritter, Ratzel), at the origin of German anthropogeography. The *Raubwirtschaft* untied a denunciation of the effects of deforestation, pollution and the destruction of resources and landscapes. It is possible that the members of the German association were influenced by this thinking, geographical and naturalistic in its foundations and marked by Darwin's evolutionary theory (Raumulin 1984; Matagne 2016).

As for the physiological bases of ecological botanical geography, they refer to Schimper. However, until the beginning of the 20th century, it could be said, as regards the contributions of physiology to the understanding of ecological phenomena, that plant geographers were often believers but not churchgoers. In their defense, we can point to a lack of training, despite the rise of plant physiology in the second half of the 19th century, particularly in Germany, where Sachs published a manual of experimental plant physiology in 1865. Physiology courses were generally linked to professional training (horticulture, agriculture, agronomy, medicine). In France, from 1896 onwards, the competitive examination (*agrégation*) offered a choice between a certificate in zoology and one in physiology. From 1898 onwards, the latter could also be chosen for a PhD (Hulin 2002, p. 110). Depending on the choices made, students could therefore miss out on training in physiology.

It was noted that the plant pathology research station in Hesse received members of the Free Association. Nothing says that the doors of the laboratories were open to them beyond a simple visit. At the beginning of the 20th century, experiments with ecological aims required instruments that were sufficiently powerful for the results obtained to be reliable. According to Ganong, author of a 1901 manual on research and teaching physiology as a basis for ecology, reconverting rustic

2 Massart is a counter-example.

equipment from old laboratories for ecological studies was adventurous (Sachs 1865; Ganong 1901, p. 7).

The Free Association had understood Warming's message, taking into account both approaches to botanical geography: floristic and ecological. Its program was developed by professionals and amateurs alike, without necessarily having a background in plant physiology or access to a laboratory and a greenhouse.

16.3. Brussels Congress

16.3.1. *High ambitions*

The 3rd International Botanical Congress in Brussels was carefully prepared. Expectations were high but the results were disappointing.

A total of 5,000 circulars in German, English and French were distributed in 1909. The official language of the Congress was French. Some 20 governments were officially represented. Continental Europe was well represented, including Russia, and also England, a number of South American countries, Persia, China and South Africa (Transvaal province).

The aim of this event was to mobilize all those, individuals and institutions, who mattered at the time in the fields of botany, phytogeography and plant ecology. Warming represented the Danish government and was also one of the Congress chairmen, along with Drude and Engler (Germany), Flahault and cryptogamist Mangin (France) and Atkinson, who was president of the Botanical Society of America from 1905 to 1907.

Warming was friends with Flahault. The latter had been to Denmark in 1890 and 1907, and returned in 1913 to take part in a botanical session organized by the International Botanical Association and directed by Warming. A Catholic, Flahault's faith was as sincere as Warming's Protestant one. They were two great figures in botany and phytogeography, both travelers to the Arctic regions, both men in the field. Like his Danish colleague, Flahault was a pedagogue, organizing excursions for his students well into old age.

Despite all the good will, "the discussion opened before the Phytogeography section was unable to lead to any definitive solution to the essential problems of nomenclature", noted Pavillard of the *Institut botanique de Montpellier*, one of the founders of the school of plant sociology (Pavillard 1912, p. 165).

Quelques congressistes ayant assisté aux délibérations de la Section de phytogéographie le vendredi 21 mai 1910 (1)

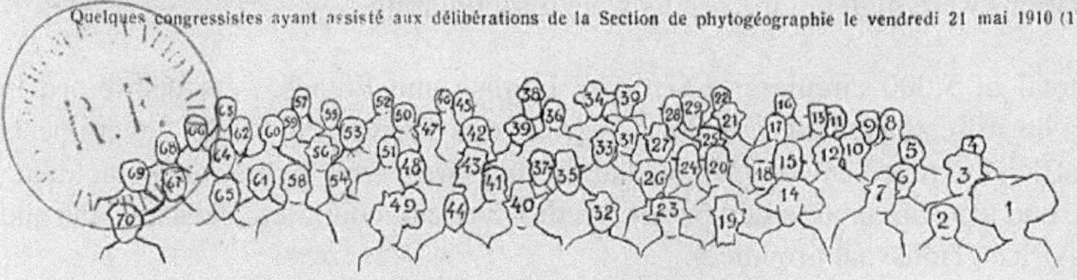

1. Madame FEDDE (Berlin).
2. M. HAROLD HAMEL SMITH (Tropical Life-Londres).
3. Prof. Dr FEDDE (Berlin).
4.
5. M. P. VAN AERDSCHOT (Bruxelles).
6. Prof. NICOTRA (Messines).
7.
8.
9. Melle X, sténographe.
10. Melle X, sténographe.
11. Madame RÜBEL (Zurich).
12. M. KNOCHE (Montpellier).
13. M. NAVEAU (Anvers).
14. Mademoiselle BODART (Bruxelles).
15. Prof. Dr ERIKSSON.
16. Prof. Dr RUEBEL (Zurich).
17.
18. Madame ERIKSSON.
19. Prof. Dr O. NORDSTEDT.
20. Prof. Dr JACCARD, Zurich.
21. Prof. GOLENKIN, Moscou.
22.
23. Mademoiselle ERNOULD (Bruxelles).
24.
25. Mademoiselle HARMS (Berlin).
26. Madame FEDTSCHENKO (Russie).
27. Prof. Dr HARMS (Berlin).
28. Prof. de DEGEN (Budapest).
29. Prof. Dr SCHRÖTER (Zurich).
30. Melle X, sténographe.
31. Prof. HAYATA (Tokyo).
32. Prof. Dr WARMING.
33. M. le Col. PRAIN (Kew-Londres).
34. Prof. BORIS FEDTSCHENKO (St Pétersbourg).

(1) Parmi les membres du Congrès de Botanique se sont glissés quelques membres du Congrès d'Agriculture tropicale qui tenait ses assises à Bruxelles à la même époque.

Figure 16.2a. *Some of the delegates who attended the proceedings of the Phytogeography Section on May 21, 1910 (De Wildeman 1910, unpaginated)*

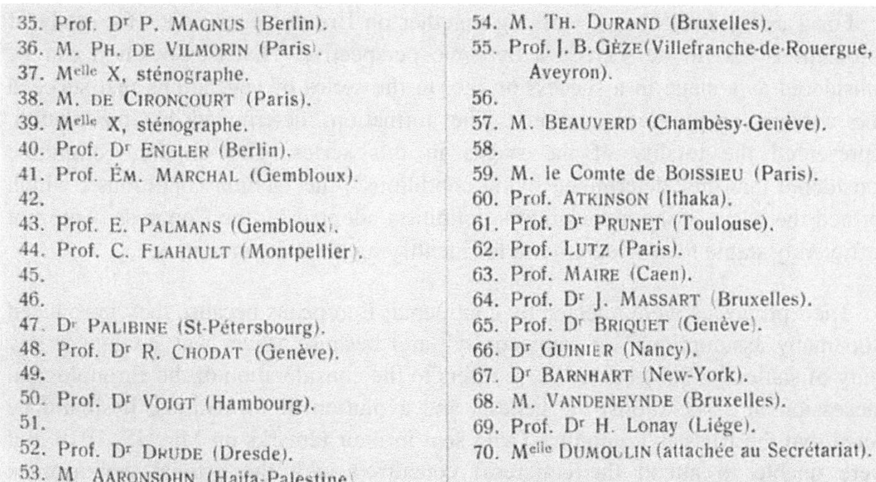

35. Prof. Dr P. MAGNUS (Berlin).
36. M. PH. DE VILMORIN (Paris).
37. Melle X, sténographe.
38. M. DE CIRONCOURT (Paris).
39. Melle X, sténographe.
40. Prof. Dr ENGLER (Berlin).
41. Prof. ÉM. MARCHAL (Gembloux).
42.
43. Prof. E. PALMANS (Gembloux).
44. Prof. C. FLAHAULT (Montpellier).
45.
46.
47. Dr PALIBINE (St-Pétersbourg).
48. Prof. Dr R. CHODAT (Genève).
49.
50. Prof. Dr VOIGT (Hambourg).
51.
52. Prof. Dr DRUDE (Dresde).
53. M AARONSOHN (Haifa-Palestine).
54. M. TH. DURAND (Bruxelles).
55. Prof. J. B. GÈZE (Villefranche-de-Rouergue, Aveyron).
56.
57. M. BEAUVERD (Chambèsy-Genève).
58.
59. M. le Comte de BOISSIEU (Paris).
60. Prof. ATKINSON (Ithaka).
61. Prof. Dr PRUNET (Toulouse).
62. Prof. LUTZ (Paris).
63. Prof. MAIRE (Caen).
64. Prof. Dr J. MASSART (Bruxelles).
65. Prof. Dr BRIQUET (Genève).
66. Dr GUINIER (Nancy).
67. Dr BARNHART (New-York).
68. M. VANDENEYNDE (Bruxelles).
69. Prof. Dr H. Lonay (Liége).
70. Melle DUMOLLIN (attachée au Secrétariat).

Figure 16.2b. *Some of the delegates who attended the proceedings of the Phytogeography Section on May 21, 1910 (De Wildeman 1910, unpaginated) (continued)*

Acot sees a number of reasons for this perceived failure, not least the fact that schools of ecology are born in different plant landscapes and give rise to specific research programs. Margalef notes that "all schools are strongly influenced by a *genius loci* that goes back to the local landscape", hindering the generalization of methods and concepts. He takes the example of "ecosystems reflecting the physical environment in which they have developed". Thus, the vegetation of the Mediterranea and Alpine regions, Scandinavian flora, the wide open spaces of North America and Russia, the tropical rainforest, do not produce the same ecology (Margalef 1968, p. 26).

In Brussels, the sense of place hovered over the debates.

16.3.2. *The break with the English committee*

The communication of the English committee's resolutions of April 24, 1909 was followed by an exchange of letters between Tansley and the English-born South African botanist Moss, on the one hand, Flahault and Schröter, on the other hand. In the wake of these long-distance discussions, the English resolutions of December 18, 1909 showed that each side approached the Congress with its positions firmly entrenched. The English emphasized their adherence to "new ideas" and "recent studies" on the succession of vegetations over time.

For Tansley and Moss – working together on British vegetation – the study of associations had to be seen in a dynamic perspective. “Each association can be considered as a stage in a succession, i.e. in the series of vegetations that succeed one another on a given station”. The formation, determined by the station, represented the totality of the stages in this series. The English committee considered that the “determined living conditions”, the “station conditions“, which formed the basis of the association’s definition adopted by the Congress, were not sufficiently stable to provide criteria for identifying plant communities.

Their proposals were rejected by continental Europeans because they introduced “too many assumptions and subjectivity” and because “there was no longer any unity of station”. “Subjectivity” here refers to the consideration of the chronological succession of associations, the genesis and evolution of formations. It should be noted that the Russian committee (who sent in their remarks on May 19, 1910, but were unable to attend the congress) concurred with the official conclusions: “existing and actual facts” had to be studied. However, they were not hostile to the introduction of a genetic principle (in the sense of studying the genesis of vegetation) (De Wildeman 1910, p. 122, pp. 134–135, pp. 143–145, pp. 163–164).

To borrow from an analogy we have already used, we could say that for the English, the association was but a transitory state fixed by photography, a snapshot whose reality was fleeting, while the succession of these images reconstituted the film in its entirety. Phytogeographers of the Old World confined themselves to the “facts” observable in photographs.

16.3.3. *The untraceable nomenclature*

The lines of research identified by the Free Association: botanical nomenclature and phytogeographic nomenclature were reflected in the organization of two sections in Brussels.

The botanical nomenclature section – some 100 participants – wrote a long report (De Wildeman 1910, pp. 43–116), showing the complexity of the task, linked in particular to the search for agreement on questions of anteriority – they determined the naming of the species attached to the name of its discoverer – and to the choice of language. The aim was to impose the use of Latin for diagnoses, which give a concise description of a taxon (species, genus, family, order, class, etc.). Thus, diagnoses fixed the information associated with the scientific name of a taxon at the time of its first valid publication.

The phytogeography section was unable to lay down rules like the botany section. It confined itself to formulating recommendations (De Wildeman 1910, pp. 117–164). The task was even more complex than in systematics, for although phytogeography was a more recent scientific field, it was sometimes based on the recognition of units that had long been named in vernacular languages.

Important preparatory work preceded the Brussels Congress, from the appeal launched in 1900 to a meeting in Zurich in September 1909. The Swiss botanist Rübel, a member of the Free Association, was commissioned to take notes for the publication of the Congress proceedings. The proposals of the various committees were also published in the first volume. Sessions were held under the presidency of Engler and the vice-presidencies of the Swiss botanist Chodat, Drude and Warming.

These preparations show that the management of the Phytogeography Section was aware of the difficulties of the task ahead. They therefore took many precautions and were cautious in having the reports of the various committees sent to them before the opening of the Congress, although this did not prevent tensions. As in diplomatic practice, it was hoped that major decisions would already have been taken before they were made public.

In the absence of consensus, it was recommended to retain the expressions used in the vernacular of each country to designate plant associations and station types. This was a question of common sense, and even of culture, as some of these units had their own local names. It was understandably difficult to generalize the use of terms that sometimes predated colonization. But this did not help to solve the recurring problems of synonymy.

For the main vegetation types, there was a reliance on Greco-Latin terms such as xerophilous or hygrophilous, proposed by phytogeographers and used in a relatively common sense. There was no despair of drafting, under the direction of an ad hoc commission, a polyglot international vocabulary giving the synonymy of phytogeographic expressions and imposing cartographic conventions, tested by Flahault in particular.

16.3.4. *Few results*

The results of the Congress included a definition of plant ecology, training and association:

– "the term ecology includes all the relationships existing between plant individuals or associations on the one hand, and the station on the other [...]. Ecology includes the study of environmental conditions and adaptations of plant species, either in isolation (autoecology), or in associations (synecology or study of formations)" (De Wildeman 1910, p. 122). We note the abandonment of the spelling "oecology" and the articulation of the term with clearly identified objects of study: plant individuals and associations in interaction, and in relation to their environment, corresponding to the station. The two approaches to ecology, self- and synecology, were validated by animal ecology;

– it is worth remembering that the notion of plant formation, initially rejected by Warming, was now used by the author of *Oecology of Plants* in 1909. Flahault, Schröter and Drude expressed their agreement with the definition borrowed from the Danish botanist: "A plant formation is the current expression of determined living conditions. It is made up of associations which are different in their floristic composition, but which correspond to similar site conditions and take on similar vegetation forms" (De Wildeman 1910, p. 123);

– "an association (*Bestandestypus*) is a plant group with a specific floristic composition, presenting a uniform physiognomy and growing under equally uniform site conditions". The German term *Bestandestypus* is translated into English as *type of stands*. It designates a type of plant community of a given floristic composition, in equilibrium with external factors. The association is the fundamental unit of synecology (De Wildeman 1910, p. 121). The definition proposed by Warming in 1909 inspired that of the 1910 Congress. The association is the ultimate expression of vital competition and adaptation to the environment, said Flahault (1900, p. 440).

Once again, we note the hybrid nature of the definition of association, which made it difficult to apply. Using the same reference, phytogeographers could produce both physiognomic and floristic analysis. The former identified the association by its physiognomy, expressed by "the relative frequency of the various forms of vegetation that make it up", and the second by its floristic composition, expressed by "the list of constituent species" (De Wildeman 1910, p. 121). The rest of the account shows that everyone had their own method and vocabulary, adding further confusion to the interpretation of the definitions of association and formation.

During a working session, the Vaud botanist Jaccard raised a real recurring question, which arose whenever an ecological unit needed to be defined. He wanted to give "association" a floristic meaning and "formation" a physiognomic meaning. His proposal, which showed a desire for clarification, was not accepted. Ever since Grisebach, there was trouble! Jaccard was unsuccessful, as the English committee disagreed with the conclusions published in the proceedings. The schools of plant sociology decided, with reference to this same Congress, in favor of a strictly floristic approach.

16.3.5. *Warming taken to task*

Professor Warming had personal issues with the English committee. Tensions were to be expected.

In the afternoon of the May 20th session, he addressed Tansley directly to respond to the English committee's "resolutions", which directly challenged his system and definitions. The representatives of the English committee criticized him for having constructed an artificial and contradictory classification. In their view, it was not growth forms (physiognomics) that could serve as a basis for classification but the total floristic composition.

In the final text of January 18, 1910, prepared for the Congress, Moss and Tansley asked that the plan to officially recommend Warming's view of plant formations be reconsidered. In the presence of the Danish professor, Tansley sought appeasement – for a change – by pointing out that Moss had only produced a summary of their work and that certain omissions or simplifications had not allowed the authors' thinking to be developed. Details of the exchange that took place during the session were not given. The transcription of Rübel's notes reports only Tansley's concluding remarks: "It was not their [Moss and Tansley] intention to make any hurtful criticism of Mr. Warming, for no one has a more sincere admiration for him" (De Wildeman 1910, p. 127).

Tansley was a great admirer of the author of the *Lehrbuch*, but not of the author of *Oecology of Plants*. We can assume that sitting in the same sessions must not have been easy for either of them.

This battle of wits was symbolic of the opposition between two visions of the ecology: continental Europeans, dominated by German and Swiss authors, and the United States, whose armed wing was the English committee. In the end, Warming's positions were supported by Engler and Drude. The section did not accept Clements' proposals, nor those of the English committee. The proceedings are unambiguous on this point.

The unanimity displayed by the participants failed to mask the reality: the English committee's disagreement on the most important recommendations was marked by the fact that there were around 14% no votes or abstentions, which is significant for this type of event.

Work in Brussels was to continue at the next Congress, scheduled for London in 1915.

16.4. A "calm" balance sheet

A few years later, in an article published in 1918 by *Annales de géographie*, Pavillard announced that he was looking back "in all serenity [on] the work completed" by the Brussels Congress. However, he was not shy about settling scores, in particular with the rapporteurs of the various committees, whom he accused a posteriori of lack of vigor. Their conclusions lacked conviction.

He denounced the "shortcomings of phytogeography" of the early 20th century, including "the inadequacy, if not absence, of any truly educational didactic tradition". Every science has a method and knowledge that must be taught. Warming could not be faulted for neglecting teaching. Pavillard was himself a pioneer in this field in Montpellier. As early as 1901, in his *Eléments de biologie végétale* (Elements of Plant Biology), intended for his students, he taught the conceptual and methodological basics that were now those of "ecological botanical geography" which involves the study of "ecological factors", as Warming encouraged (Pavillard 1901, p. 545, 1918, p. 403). Moreover, after the Free Association, he insisted on the need to articulate floristic knowledge and "phytogeographical speculations", lest people "lost their way in abstraction". Nevertheless, Pavillard thanked the Brussels congress for proposing a definition of association that allowed it to be seen as the result of the "survival of the fittest".

In a period dominated by the floristic approach to plant ecology, he defended that the reality of the association's existence was expressed by its floristic contingent: "The association is established and maintained in a community whose members are the plant species, the fundamental and only concrete units of systematics [...]. The exact and complete inventory [of its floristic procession] imposes itself on the phytogeographer". He thus distanced himself from the definition validated by the Brussels congress, which, as we have seen, juxtaposed the floristic and physiognomic approaches.

At first, Warming was spared by his critics. Furthermore, while Pavillard welcomed the contributions of Drude and Engler at the Brussels congress, it was the Danish botanist who received his most fervent praise. Warming could claim authorship of the proposed definitions, in a language of great clarity. Thanks to him, "in ecology we distinguish three fundamental units, whose gradation is [...] in descending order: station, formation, association". "Under the energetic impetus of Mr. Warming, a new notion, of at least equal importance [to that of station], arose in the ecological field: the notion of growth form or vegetation form" which is based on the notion of epharmony (Pavillard 1918, p. 403, p. 405, p. 407).

However, in the final pages of this 1918 article, Pavillard, in the manner of Tansley, delivered the final blow:

> The ecological thesis of M. Warming, grafted on the physiognomic tradition of Grisebach and amalgamated with the floristic conception of associations of M. Flahault, gave the most indigestible scientific broth, of which the congress of Brussels, fortunately, did not dare sanction the formula definitively. We had a narrow escape, but the situation remained just as confused, the imbroglio as apparently inextricable (Pavillard 1918, pp. 412–413).

A quarter of a century after the Brussels congress, in *Éléments de sociologie végétale* (Elements of Plant Sociology), appeased, he thanked Warming. Thanks to him, "the ecological meaning [of the station] is obvious" (Pavillard 1935, p. 6, p. 15).

Tansley recognized in 1947, at a time when the foundations of systems ecology had been laid, that Warming, along with Schimper, was the founder of modern ecology. In the meantime, in 1920, the British botanist visited Warming. In this context, and despite his reservations about *Oecology of Plants*, he testified to his unfailing admiration for the pioneer of ecology. He was even touched by the 79-year-old man's enthusiasm for collecting plants, impressed by the acuity with which he pointed out the different characteristics of life forms.

17

The Man of Knowledge and Power

Warming had been building his career since the late 1860s. He took part in Scandinavian and German scientific meetings from 1868 until 1916. The International Congress of Botanists, Horticulturists, Traders and Manufacturers of the Plant Kingdom in Amsterdam, attended by 500 people on the premises of the Dutch Zoological Society from April 13 to 17, 1877, was an opportunity for him to publicize his research on *Cycadophyta* ovules (gymnosperms that look like palm trees but grow in thickness). He took part as vice president. At the time, he was a docent at the University of Copenhagen.

He had long been visiting foreign universities: Strasbourg and Paris in 1876 and 1880, Göttingen, Jena, Bonn in 1880, etc. He made a name for himself by taking part in major congresses as early as 1877 (Amsterdam) and became president of the International Botany Association (1913). He was invited to the celebration of Linnaeus in 1907 in Uppsala for the bicentenary of the great Swedish naturalist's birth and on July 1, 1908 for the 50th anniversary of the Linnaean Society of London, during which extracts from Darwin's work were read, a letter from him to the American botanist Asa Gray, Wallace's manuscript. One of the aims of the meeting was to reaffirm Darwin's intellectual priority (Darwin and Wallace 1858; Gayon 1992). On this occasion, Strasburger and Haeckel received the Darwin–Wallace silver medal.

Warming also received prestigious honors and medals: *rector magnificus* from the University of Copenhagen (1907–1908), 1st degree Commander of the Order of Dannebrog (a Danish honorary order created in the 18th century), the Royal Victorian Order (honorary order of the United Kingdom and Commonwealth created by the Queen), the Brazilian Imperial Order of the Rose (for services to the State), the Erzherzog Rainer Medal[1] of the Imperial Royal Society of Zoology and Botany of Vienna (1911) and the Grand Gold Medal of the Royal Swedish Academy of Sciences (1922). Some of his anniversaries were officially celebrated, such as his 70th birthday. Hanne had hoped that her husband's 60th birthday would already have been celebrated with a grand party, inviting numerous botanists, but she wrote: "His modesty forbade it". However, he could not escape the visits of colleagues, who came one after the other to his home (Kolderup Rosenvinge 1911; Prytz 1984, p. 133).

17.1. A pillar of the Carlsberg Foundation

At the University of Copenhagen, Warming shared the scientific, political and religious views of the influential Japetus Steenstrup. He always supported his former zoology professor. In retrospect, he had no regrets about having chosen the wrong side in the early 1870s, taking up the cause of Japetus, which had temporarily cost him his position at the university, as we have seen. A long friendship linked him to Johannes, son of Professor Steenstrup, who had become a conservative historian, in tune with his father and his friend Warming on political and religious matters. Warming joined Steenstrup on the board of the Carlsberg Foundation. For 32 years, he was one of the Foundation's decision-makers and a strong supporter of Danish science. Finance being the lifeblood of research!

1889 was an important year for the Carlsberg Foundation and for Warming. The historian Holm and Steenstrup were re-elected to the Board of Directors for a further 10 years and Warming joined the Board. The creator of *Carlsbergfondet* (1876), Jacob Christian Jacobsen, dedicated some of his shares in the Carlsberg brewery to the laboratory (1875) for fermentation physiology and technology and to the National History Museum at Frederiksborg Castle. He donated 1 million Danish kroner to the Foundation, which at the time represented a huge sum (Poulsen 2016, p. 45). The two main functions of the powerful Foundation were to support the laboratory and Danish scientific research.

Warming served assiduously on the Board until 1921, when he resigned due to age and deafness[2]. During his long tenure at the Foundation, his institutional and personal authority gave him considerable influence on all national issues relating to natural history.

1 From Archduke Rainer, great collector of ancient papyri. Curator of the Academy of Sciences and patron of the Austrian Museum of Art and Industry.

2 For the year 1922–1923, the members of the Board of Directors were Drachmann, Erslev, Henriques, Hjelmslev and Ostenfeld.

Figure 17.1. *The Carlsberg laboratory and the statue of its founder Jacob Christian Jacobsen (public domain via Wikimedia Commons). For a color version of this figure, see www.iste.co.uk/matagne/birth.zip*

He also worked alongside Steenstrup at the Royal Danish Academy of Sciences and Letters, supported by the Carlsberg Foundation. The Academy received 10,000 kroner each year "to ensure its independence and freedom" from the Carlsberg Foundation. In 1899, it was officially installed in new premises built by the Foundation, in the presence of the King and several royal princes. Historian Holm was Chairman of the Board (Royal Danish Academy of Sciences and Letters 1896, p. 17, 1898–1899, p. 12). A nationalist Christian, Holm won respect for his decisions, independent of his political or religious sympathies or antipathies.

Each year, the Academy published a report on the scientific activities supported by the Foundation, with details of the accounts and a breakdown of support, grants and funding granted to various projects: missions, expeditions, publications (natural history, literature, archaeology, astronomy, mathematics, etc.), botanical gardens, etc. Jacobsen created the National History Museum at Frederiksborg Castle; Holm was a member of the administrative board. From the 1890s onwards, the multidisciplinary management of the Foundation also included Jørgensen (fermentation physiologist, who was Warming's assistant), Johan Louis Ussing (archaeologist and epigrapher, he launched the first Danish archaeological research in the Mediterranean and the excavations of Rhodes) and Zeuthen (mathematician and historian of mathematics, member of the Academy of Sciences and Letters). They promoted their fields and their own work. Thus, many Warming research projects benefitted from the support of the Foundation (on Greenland, the Faroe Islands, Iceland, the flora of Denmark, etc.).

Members were re-elected for a further 10 years. Steenstrup died in 1897 and was replaced by physicist Christian Christiansen (supervisor of Niels Bohr who had received a Carlsberg Foundation grant) for the remaining three years. Warming appears to have set a record for longevity at the Carlsberg Foundation. In 1921, he was replaced by one of his former students, Hansen Ostenfeld, who had become a botanist (he had been part of the mission to the Faroe Islands) and was also elected for 10 years. The Foundation was thus characterized by the stability of its directors and by a form of "social grouping", which ensured continuity in its scientific policy. Its executives were to be found in museums and academic societies, at universities, as editors of scientific publications and, in some cases, in the corridors of power.

At the Foundation, Warming felt he could work for the good and honor of his country (Kolderup Rosenvinge 1924, p. 126). His choices in financing research benefited Denmark. In 1897, Børgesen and Paulsen, zealous students, were awarded 1,200 crowns for an illustrated publication on the vegetation of Denmark's western islands. Another of his students, Schmidt, whose abilities he had recognized at an early stage, was consistently supported by Warming in all circumstances. It had even been said that Warming "fought like a lion" for his student, and was able to convince his colleagues at the Foundation to give him their support (Poulsen 2016, p. 48, p. 130, p. 197, p. 144). Finally, he obtained a grant for him to study the flora of Siam's coastal areas. He was convinced of the importance of his protégé embarking on his first scientific expedition. He was not mistaken. It was a success and a springboard for Schmidt's career. Dividing his time between botany and marine zoology, Schmidt was appointed head of the physiology department at the Carlsberg Foundation laboratory, replacing Hansen (1909). Marine biology became his main activity.

17.2. A Lutheran conservative and a patriot

In the 19th century, Denmark went through crises and wars: first battle of Copenhagen in 1801 against the English; second battle in 1807, the English bombarded the city (in retaliation for Denmark's alliance with France during the Napoleonic Wars); the government declared the country bankrupt (1813) and found itself relatively isolated, while Sweden dominated Norway (lost to Denmark) and Finland; the two painful Duchy Wars reduced the Crown's territory and put Danophiles and Germanophiles in conflict; the sale of colonies to the English weakened the Danish Empire.

After an 18th century of prosperity, these trials and wars were particularly traumatic for Danish identity. The period was, however, a golden age of artistic, scientific and literary achievement, in a period of contraction and retrenchment. The great names of the arts and letters were Andersen, Kierkegaard, the painter

Eckersberg (historical paintings, portraits and landscapes) and his pupil Købke, the sculptor Thorvaldsen and composer Gade. Ørsted discovered the interactions between electricity and magnetism, the origin of electromagnetism.

King Frederick VI, liberal at the start of his reign, embodied a form of conservatism that went hand in hand with Denmark's withdrawal into itself. However, the first Constitution of 1849 (which did not apply to Schleswig because of the Prussian invasion), amended in 1853 and 1866, was more liberal than in Sweden, and created the conditions for a parliamentary regime that was strengthened by the three successive reigns of Christian IX, Frederick VIII and Christian X. This regime really took hold at the beginning of the 20th century, laying the foundations for the present-day state. Collective freedoms were guaranteed. Strong executive power was entrusted to the King, legislative power shared between the Crown and the parliament (*Rigsdag*), composed of two chambers.

Denmark entered the era of modern capitalism late and its industrialization was slow (it had no coal). On the contrary, it could rely on its agriculture, which, in crisis in the 1860s–1880s, underwent measures and modernization. This involved the sale of unprofitable farms, conversion to intensive livestock farming (cattle, pigs, sheep, poultry) and useful crops linked to the food industry (sugar mills, dairies, canneries, breweries), rational exploitation of forests (cellulose, paper pulp), drainage and reforestation of the Jutland moors and modernization of the fishing industry. "Denmark became a huge breeding farm, especially for cattle and pigs" (Mougel 2006, p. 17). The economy depended heavily on exports to the United Kingdom. In 1900, 75% of the population was rural, 60% in 1914, with industry accounting for a quarter of jobs. Copenhagen then had a population of 570,000. After emigration linked to the political and economic situation and to insecurity, the demography became positive (1.75 million inhabitants in 1875, 2.75 million in 1914). Many families were large, including in the upper echelons of society, but infant mortality remained very high.

Neutral during World War I like Norway and Sweden, or more accurately non-committed, Denmark continued to maintain commercial relations with everyone. However, it was most affected by its shared border with the *Reich*. It was forced to mine the Sund and Kattegat straits, which controlled the entrance to the Baltic Sea, while the country was more on the side of the Entente (Italy, France, British Empire, Russian Empire until 1917, United States in 1917) (Fol 1978; Helle 1992, p. 113 *ff*). With the intensification of submarine warfare, the Danish fleet lost a quarter of its tonnage. Against a backdrop of crisis (200% inflation, food shortages, strikes and riots), Denmark sold the Virgin Islands to the United States for $25 million, but retained full control over Greenland.

The question of whether part of Greenland belonged to Denmark was one of the issues at stake during the Danish expedition of Ludvig Mylius Erichsen (1906–1908). The aim was to refute the idea that part of northern Greenland belonged to the United States, based on the supposed existence of a separation of the territory by a strait, the Peary Channel, according to erroneous cartographic surveys (expedition by American Robert Edwin Peary 1893–1909). Erichsen traveled southwest along what turned out to be a dead end, a fjord (later named Denmark Fjord), preventing the Danish from considering that the northern part of Greenland could be legitimately claimed by the United States. It was another expedition, led by Ejner Mikkelsen (1909–1912), who had gone in search of Erichsen, who had disappeared in 1907, that led to the recovery of the cartographic documents left in a cairn (a pile of stone protecting the precious documents) at the head of the Danish fjord. Erichsen wrote that the Peary Canal did not exist. Mikkelsen told the story of his perilous three-year expedition in 1913 (*Two against the Ice*), and an Icelando-Danish film from 2022 is based on it (*Against the Ice*) (Mikkelsen 1913)[3]. It bears witness to Danish explorers' centuries-long fascination with Greenland, a territory coveted by several US presidents (George Washington, Harry Truman, Donald Trump).

In 1908, a moving national tribute was paid to Erichsen at the University of Copenhagen. At the podium, Warming evoked the memory of the expedition's departure and associated Erichsen's courage with that of the Vikings who had fought through the immensity of the polar night and ice. Greenland explorers, including Warming, worked for science and Denmark. Then, a song by writer and publisher Nielsen continued the ceremony. One stanza in particular struck a chord with the audience. It celebrated those who had fallen in the name of Denmark. The ceremony ended with a song by Møller, whose words were composed by Joannes Helms, *Jeg elsker de grønne lunde* (I love green groves). They were written for one of the open-air meetings at which political issues relating to the Constitution of June 5, 1873 were discussed. Helms, a writer and teacher, was involved in the second Duchy War. His song celebrated Danish nature, loyalty and love of country[4]. The painters of the Danish Golden Age, 1810–1840, exalted the landscape and national nature. In other words, a sense of nature and patriotism went hand in hand. For the Danes, their landscapes were beech forests, fragrant hedgerows and fertile fields, on which Warming also cast his eye as a sensitive naturalist and phytogeographer (Delavigne 2003, p. 205).

3 *Against the Ice*, film by Coster-Waldau and Derrick, directed by Finth, 2022, filmed in Greenland and Iceland.

4 "Jeg elsker de grønne lunde" ("I love green groves") was featured in the Danish TV series *Matador* (directed by Balling), broadcast between 1978 and 1981.

Other events bore witness to a patriotism exacerbated by the geopolitical situation. For example, the awarding of the Nobel Prize in 1917 was subject to war-related tensions. The "cohesion of neutrals" (Denmark, Norway, Sweden) was put to the test. The award of the prize to Karl Gjellerup, a pro-German (he translated several of his works into Goethe's language), supported by Germanophile Sweden – the University of Uppsala was described as a "suburb of Berlin" – was tempered by the joint attribution to Henrik Pontoppidan: "The scandal was halved. Pontoppidan, trained in the school of naturalism and realism, was "deeply rooted in the popular and religious traditions of his country" (Jolivet 1961, p. 10, p. 12, pp. 24–26). His novels painted contemporary Danish life and captured the political, social and religious transformations from the second half of the 19th century to the 1920s. Pontoppidan was called the Danish Balzac.

While a student at Copenhagen Polytechnic, where Professor Warming was a visiting lecturer, Pontoppidan applied for a place on the Greenland expedition planned for 1884. He prepared to go, but in the end was not chosen. After this rejection, he wrote a short story, *The Polar Bear*, set in Greenland.

Between Warming's departure for Brazil and his return, the Second Duchy War sealed the fate of the southern Jutland territories for a long time to come. Thanks to the letters sent by his uncle, he had been able to follow the events, which did not spare Kolding, with a time lag of five or six weeks. Permanently affected, Warming campaigned for the reunification of the lands lost by Denmark. He became openly anti-Prussian, a national conservative explicitly hostile to socialism and, of course, to anarchism, whose rise he feared, while the conservative party was in relative retreat at the end of the century, while social-democratic parties, on the German model, were gaining strength. He expressed his concern in a letter to his son Jens (a professor of economics) on September 15, 1898 (Prytz 1984, pp. 9–11, p. 117). In fact, in 1901, the Right collapsed and parliamentarianism was established in favor of the Lower House. At the beginning of the 20th century, Denmark was working to democratize its institutions, social progress and its national security.

Warming's commitment took the form of his financial contribution to a secret fund, to help farmers in Jutland who were defending Danish culture and spirit by fighting the Germanization of northern Schleswig to acquire farms. In a letter to his son Jens dated December 7, 1899, he referred to the "October 5th Association", whose existence was unknown to the Germans. It was founded on October 5, 1898 by 45 people gathered in Copenhagen. It managed funds that enabled the acquisition of 39 properties in Nordslesvig (Northern Schleswig), thus preserving land in Danish hands, wrote Warming. The association survived on donations and legacies, public appeals were obviously excluded (Petersen 1972).

An active Protestant, involved as we have seen with students defending the evangelical faith against Darwinism, Warming was an authentic representative of a Lutheranism that "everywhere imposed a strong religiosity coupled with a rigorous moralism" (Mougel 2006, p. 13, p. 25). He was convinced that religion contributed to social progress and education. Finally, he put his scientific expertise at the service of his homeland. His in-depth knowledge of plant communities and their ecology enabled him to draw up an inventory of natural resources and provide information on how to make the most of them.

According to Oluf Friis, a literary historian, the Jutland temperament had particular characteristics: "the legal head, the clear mind that trained scholars, pride and the instinct for independence" (Friis 1929, p. 88). One may be tempted to apply this formula – seductive and reductive as formulas often are – to the Jutlander Warming. The "legal head" was that of the "scholar" who had always been able to defend his rights by making decisions that were calculated and integrated into a legal framework, as well as those of his country, with clarity, consistency and pugnacity. He was a Jutlander who cultivated pride in being Danish and its independence. These latter aspects connected him more broadly to Scandinavian culture.

17.3. Scandinavianism

Northern Europe, north of 54th north parallel, has a quarter of its territory beyond the Arctic Circle. Its coasts are bathed by three seas: the Norwegian Sea, the Baltic Sea and the icy Arctic Ocean.

Warming was a Danish patriot who belonged to the Scandinavian cultural area (Sweden, Denmark, Norway, stricto sensu)[5]. Thanks to his networks, travels and scientific expeditions, he was well acquainted with the northern kingdoms, where he built up a network of friends and scholars. He even lived and taught in Sweden for three years.

Despite crises and wars – and sometimes thanks to them – a certain unity was achieved around the values and ideals of Scandinavianism, a literature that developed the theme of the "genius of the North", and also around a monetary union (1873) and a common currency (1875), the *krone*.

---

5 It is distinct from the Scandinavian Peninsula (Sweden, Norway) and Fennoscandia (Scandinavian Peninsula, Finland, Karelia, Kola Peninsula).

Scandinavianism is characterized by:

> A cult of community and independence, a taste for nature and freedom, a civilization founded on a deep attachment to equality, civic-mindedness, pragmatism and order, an innate sense of what separates neighboring peoples – Germans, Slavs and Balts – and a recurring aspiration to unity, from the Scandinavian empires in the high Middle Ages to the Scandinavianism of the 19th century, via the Kalmar Union [Denmark, Sweden, Norway], which would bring the peoples of the North together into a single entity from 1397 to 1523 (Mougel 2006, p. 6, p. 28).

After Sweden, Denmark fortified its Viking roots, especially from the late 18th century onwards. A historical school nurtured a literature that blended fact and myth, in the service of a Scandinavian quest for identity. Danish Cimbrism (the Cimbres are a people originally from Jutland) and Gothism (in reference to the ancient Gots or Goths) in Northern Europe offered "literary material, morals, wisdom and even metaphysics" for a long time to come. For historians, the Treaty of Kiel in 1814 also marked "the Nordic revival" identified with the mythical figure of the Northern hero facing perilous adventures in defiance of danger (Durand 1957, pp. 10–11).

Northern Europe was built on a dual heritage, Viking and Lutheran. The culture of Denmark, between continental and Nordic Europe, is understood in this dual context, enriched by the warlike encounter of Viking culture with Christianity and the Latin language, although Danish political ambitions turned northwards from the Middle Ages onwards. Two events had a profound impact on Danish history: the Lutheran Reformation under Christian III (1536) and the advent of absolute monarchy (1660–1849).

Scandinavism in the 19th century, also known as pan-Scandinavism or nordism (extended to Iceland and Finland), is understood on cultural, political and literary levels. Denmark, "which enjoyed its literary golden age until the 1860s, was a vector of Scandinavian influence in the cultured circles of the Norwegian capital" (Largùèche 2015, Chapter VI, p. 119). Scandinavian events in Christiania, Copenhagen, Lund or Uppsala, until the 1860s, supported a movement to rediscover national literary treasures, launched by historians and philologists. In Norway:

> The Scandinavian meetings of 1851–1852 were the occasion for sumptuous parties at which dozens of poems and songs were declaimed. More generally, a veritable student folklore developed during this decade, made up of ballads and light comedies, declaimed or performed at various festivities (Larguèche 2015, Chapter VI, p. 23).

In both Christiania and Copenhagen, students, academics and the capital's good society joined in parades, political discussions, receptions, theatrical or musical performances, balls and country walks. The first general scandinavist meeting was held in Copenhagen in 1845, the second in Uppsala in 1856 and the last in Christiania in 1896.

In 1845, around 150 students from Uppsala were joined by those from Christiania and Lund in Malmö. In Copenhagen harbor, they were greeted by the Danes with a poem by the great storyteller Hans Christian Andersen, who, after a visit to Sweden, wrote "I am Scandinavian" (*Jeg er en skandinav*). It celebrates "the beauty of the Nordic spirit, the way in which the three sister nations [Sweden, Denmark, Norway] have gradually grown together". Swedish composer Otto Lindblad set the poem to music, to Andersen's great satisfaction. "The young Scandinavians were lavishly welcomed to Copenhagen on Midsummer's Eve. The festivities lasted until Friday June 27".

The arrangement by Danish composer Johan Christian Gebauer was less successful. The melody, more difficult, was described as "new music". Nevertheless, it was sung at meetings, performed by a men's choir at the Casino Theater in 1848, at political gatherings and included in several songbooks until 1870:

> The Scandinavian festival was not, however, initiated as a popular festival: it appeared as a rather traditional, relatively regulated festival, but its ritual allowed for the impressive symbolic occupation of an area between Christiania, Copenhagen and Uppsala. In the cities, it was celebrated in places such as royal buildings. The social formalism of the Scandinavian moment, the respectful homage to the reigning sovereigns and the constituted bodies were not incompatible with the desire to mark the passage of the "Young Scandinavians" into an extended and open geographical space, in the cities and in the regions crossed (Largueche 2015, Chapter VI, p. 114).

In political terms, Scandinavianism aimed to turn former enemies into allies, against external enemies. It was a movement that defended the idea of a unified Scandinavia, founded on the promotion of common cultural and historical bases, on a shared mythology and linguistic roots (Old Norse).

The movement did not recover from the Duchy War, and *Fædrelandet* (The Fatherland) changed its political line and became a conservative newspaper, having previously been a spokesman for the Danish democratic cause. Scandinavian Danophiles saw the defection of Sweden and Norway as a betrayal. After World War I and the restitution of lost territories, political Nordism, described as a form of "collaborative nationalism", developed, and is then structured around the Nordic Council in the 1950s (Sweden, Norway, Iceland, Denmark).

For Warming, being a religious patriot in this context meant "supporting the authority of the King, religious and social stability and the Danish imperial ambitions". While scientific, his "community concept also drew its inspiration from the Danish political and social environment". "It provided a tool for managing nature" (Anker 2011, pp. 325–327).

18

The Protagonist of the Danish Ecological Project

After his return from Stockholm, Warming could focus his research on his homeland and the territories controlled by Denmark. He made no secret of his political commitment and put his skills in botany, botanical geography and plant ecology to the service of the kingdom and the Danish colonial empire.

18.1. Denmark's scientific prestige

There is a tradition that values and disseminates the scientific excellence of Denmark and Scandinavia, at least since the 18th century for botany.

An encyclopedic botanical atlas, *Flora Danica*, was published from 1761 to 1883 on the initiative of Oeder, director of the Copenhagen Botanical Gardens. He had been summoned by the King, but the academics refused to admit a foreigner – he was German – to the teaching staff. He was finally appointed *professor botanicus regius* (royal professor). *Flora Danica* dealt with the flora of areas within the Danish Kingdom.

Several renowned naturalists succeeded one another as the publication's editors over the period: Friedrich Müller, Vahl, Wilken Horneman, Drejer (friend of Steenstrup), Schouw, Vahl (Martin's son), Liebmann, Steenstrup and Lange. A total of 51 volumes were published. The flora was a great success with both professional and amateur naturalists. In the mid-19th century, with the rise of Scandinavianism, the Society of Scandinavian Naturalists proposed to include plants from Norway and Swedish endemics in three supplements from Sweden.

For political reasons, flora inspired Crown Prince Frederick to commission a porcelain service decorated with "Flora botanica" as a gift for the Russian Empress Catherine the Great (Catherine II), who died before the service was completed. The service faithfully reproduced plants from Oeder's atlas, and indicated the Latin names of the species represented. Completed in 1802, the 100-person service was presented to the royal family. What remains of it can be seen today at Rosenbourg Castle (Wojciech 1987).

From the end of the 1870s, through his review of authors in the field of botany and phytogeography, Warming was in line with those who worked to make Danish flora known, both inside and outside the country. He considered it necessary to pay homage to ancient botanists and to do them justice by mentioning their work, deploring the fact that they were sometimes relied on without being cited. All of his work was marked by great vigilance on this point. He added that Danish and foreign botanists who had published in Denmark since the 16th century in horticulture, forestry, agriculture and animal husbandry also had a vocation to appear in his lists. This was not trivial, as they had special knowledge and expertise in the management of agricultural and forest land. Warming appealed to those who wanted to make themselves known and wished to contribute to the project by providing revisions and additions. This work was also undertaken and continued by one of his former students, Christensen (Warming 1880–1881; Christensen and Hansen 1913–1963; Christensen 1924–1926).

18.2. Fundamental and applied ecology

Warming's research was at the service of an ecological, economic and political project that involved understanding and developing the kingdom's natural resources. They included the question of human intervention, a recurrent theme at least since his stay in Brazil.

18.2.1. *The impact of human activities*

In *Lagoa Santa,* he did not pass negative judgment on local uses of the land, predating colonization and continuing beyond. Anecdotally, he even noted that *derrubadas*, the cutting of forests to create farmland, facilitated the work of the botanist, who could thus conveniently determine which species had fallen to the ground (Warming 1892b, p. 141). Nor did he judge the practices of nature in Greenland or the Arctic regions in general, but only measured their floristic effects and their impact on plant formations.

Figure 18.1. *A derrubada: on the hill behind the fazenda (landholding), the forest is felled (Warming 1892, p. 242)*

On the contrary, in *Lehrbuch*, he gave his opinion on methods of managing natural environments in Denmark, the forest in particular:

> Jutland [especially eastern Jutland, the island of Seeland] was once covered with oak forests [the post-glacial climax forest]. Leaching of the upper soil layers by precipitation must have reached a degree high enough to prevent forest regeneration, and *Calluna* heath [heather] took its place. Then other factors came into play [...]. Careless and ignorant logging, the use of Jutland timber in the Middle Ages, large-scale mining of iron ore and westerly winds swept the forest away (Warming 1896, p. 369, 1902, p. 384).

He blamed "poor forest management", which led to the clearing of forests. Over a long period, the ecological advantage was given to beech and then to heathland.

In Chapter 4 (Warming 1996, pp. 370–372, 1902, pp. 385–387), he shows that the ecological status of heather moorland can differ according to historical and ecological contexts. Where a long period of forest establishment has allowed the heather to reach a point of equilibrium, its destruction leads to the victory of the heather, which colonizes the newly open environment, along with other species. In this case, the heath becomes a "form of semi-culture" (*halbkulturform*) (Warming 1896, p. 372, 1902, p. 387). This is undoubtedly how the changes in Russia and northern Germany can be explained.

In other regions, heather heath is the natural end-stage, as on the poor sandy soils of western Jutland, which have not been disturbed and have not given rise to forest formations before being cleared and amended. Danish poet Steen Steensen Blicher was inspired by this "brown sea of heather". In the fall, the hills "take on the purple glow of heather, which is only just beginning to bloom" (Wyss-Neel 1971, pp. 22–23). A two-page chapter in *Oecology of Plants* discusses the influence of grazing and cultivation on flora and vegetation and the composition of plant associations at different latitudes (Warming 1909, pp. 326–327).

Warming seemed unaware of the pioneering work of American diplomat and philosopher George Perkins Marsh, who, before him (1864), denounced the disastrous management of the forest, in a perspective that was to become that of the first environmentalists. Yet Warming's denunciation of "forest mismanagement" used the same words. Marsh's argument was taken up at the beginning of the 20th century by President Theodore Roosevelt, who advocated a reasoned exploitation of nature (Marsh 1864).

In Europe, in the 19th century, deforestation intensified with the extension of agriculture and livestock farming and the use of timber and firewood. A blast furnace burned the equivalent of 50 hectares of forest a year. With the proliferation of coke ovens, wood resources were better preserved but landscapes were severely impacted. In the 1850s, "miners' colonies" housed thousands of workers in the Ruhr Valley. Deteriorating sanitary conditions and air and water pollution were the price to pay. Far from it, industrialization did not reach the same proportions as in the German Empire. The Danes were "a people who have enough to do with damming themselves against the sea and protecting their fields against quicksand", wrote historian Louis Paul-Dubois in 1909 (Carlsen et al. 1900, p. 65; Paul-Dubois 1909, p. 660). Until World War I, the Scandinavian countries shared this essentially agricultural character, oriented towards silviculture and fishing.

Warming pointed out the impact of major works on vegetation (drainage, damming, hydraulic equipment, agriculture, livestock farming). The cultivation of the moors in the 1880s and 1890s by small farmers (*husmaend*) was encouraged by the Moors Society, founded in Aarhus in 1866, which dug 400 km of irrigation canals (Schou 1900). These irrigation issues are addressed by Pontoppidan, through his character in the novel *Lucky Per* (*Lykke-Per*).

Denmark at the time was a country of smallholders, whose economic wealth stemmed from the production and export of livestock products. A network of dairies on the mainland and islands transported milk by land (aided by the development of the railroads) and by ship. From 1875, the new port of Esbjerg transported meat (especially pork), butter and eggs three times a week to the London market and the industrial districts of northern England. Farm animals were also exported to Germany.

Warming insisted on one indicator of good soil health: the presence of earthworms. He returned to this point several times, sometimes including other burrowing animals. The ecological role of earthworms was recognized as early as the 18th century by British naturalist Gilbert White. In his *The Natural History and Antiquities of Selborne* – a small area nestled between the Hampshire hills – he observed nature and discovered that "Earth worms, though in appearance a small and despicable link in the chain of nature, yet, if lost, would make a lamentable chasm […]. Worms seem to be the great promoters of vegetation" (White 1789, p. 216). Warming quoted Darwin, who took up this idea of the role of earthworms (Darwin 1881).

We can today confirm the essential ecological functions of these animals. They dig tunnels that allow water and air to penetrate the soil. This infiltration limits runoff and erosion and brings water into contact with plant roots, which grow more easily in soil aerated by the numerous tunnels. As they feed, earthworms recycle organic matter by participating in its decomposition. After mineralization by a host of different organisms, the organic matter is once again available to plants.

18.2.2. *Ecological dynamics*

In 1904, the Royal Danish Academy of Sciences and Letters published *Bidrag til Vadernes, Sandenes og Marskens Naturhistorie* (Contribution to the natural history of the Wadden Sea, sands and marshes), an ecological study in which Warming considered the flora and fauna on a substrate dominated by sand or clay. For the fauna, he called on several contributors, in particular Wesenberg-Lund, a zoologist and pioneer of limnology and nature conservation in Denmark. He fought against raptor hunting, uniform forestry and water pollution. Wesenberg-Lund was supported by the Carlsberg Foundation and was a close associate of Steenstrup and Warming. Like Warming, Wesenberg-Lund was sympathetic to Lamarckism and his non-dogmatic, religious worldview marked his descriptions of the harmonies of nature.

In this study, Warming identified two main zones, with transitions between them: the arenicolous zone and the corophilous zone. The first is easily identifiable by the multitude of worm castings to the surface by worms buried in the sand. The second gives its name to the small crustaceans that, like the arenicoles, can form very dense populations. They also build burrows.

According to Warming, fauna, much more than flora, was responsible for the formation of these environments. On the contrary, in muddy terrain, plants are active in fixing the mud and transforming it into salt meadows, although gastropods play a significant role. He identified intermediate formations between sandy and clayey mud. Photographs taken in Holstein illustrate these in the publication.

Ecological dynamics was studied by Warming. Through case studies, he described the natural and anthropogenic processes impacting plant formations that are destined to be, or already have been, exploited.

18.2.3. *Case studies*

Among Warming's many field studies in Denmark, two in particular illustrate the way in which he integrated the natural and man-made dynamics of the environments he studied.

At the north-western tip of the island of Fanø – today the northernmost of the Danish Wadden Sea islands – or on the peninsula and small peninsula of Skallingen and in several coastal areas of the North Sea, there is a sandy plain, prone to flooding during particular weather events, such as storms. Flowering plants thrive in these environments. On the landward side are dunes, then "arenaceous meadows" (green meadows more or less invaded by sand). The surface sand is fixed by a layer of green algae (Phycochromaceae) a few millimeters thick, which forms a fairly firm soil. This zone of arenophilous algae is populated by a small association of gallery-digging Coleoptera (*Bledius* predominate), and nematodes are common. Many insects (Hymenoptera) hibernate probably in the dunes. Diatoms, microscopic algae at the base of food chains, are numerous. They were studied by another collaborator of Warming, Ernst Vilhelm Østrup. Food chains start with algae, followed by phytophages (which, in this case, consume algae), themselves consumed by zoophagous animals (Warming 1904b, 1906–1919).

Warming knew this part of the coast and its islands very well, having explored it several times personally and with his students. At the time, he was just a stone's throw from his native island, to the south. Over the years, he gained a pretty clear idea of how these environments worked. In 1913, he was the best guide, very alert despite being "three times twenty and ten", that the International Association of Botanists could find for its annual meeting. It was officially received by the Danish Botanical Society. Twenty-five excursionists, Norwegians and French, strongly represented, as well as Dutch, Swiss, Russians, Japanese, Americans and British, were more interested in the ecology of these particular environments than in the relatively poor flora, compared to that of the tropics or the Mediterranean. Charles Flahault, whose presence was noticed, was not the least bit interested.

They were not wrong. The Wadden Sea offers an exceptional ecosystem and the largest intertidal zone in the world. The tide uncovers vast marshy areas, forming habitats for flora and fauna (channels, sandbanks, mudflats, marshes, estuaries, beaches, dunes, mussel beds). Five thousand animal and plant species have been recorded, representing a unique biodiversity in such a northerly zone.

Excursionists were taken to other environments over the course of a session lasting a dozen days. The atmosphere was studious, warm and relaxed, to the extent that Professor Flahault gave a "little song" (in French, in the text of the minutes) and Professor Hasberg Gran, a Norwegian specialist in phytoplankton, sang a folk song (Smith 1914, p. 65, p. 70).

The island of Fanø allowed Warming to model the ecological dynamics of interest to his audience. Between the hotels and the north-western tip of the island, a first formation, that of the dunes, was emerging. Tussocks of tall grasses offered protection to more delicate plants, whose density was increasing. In the hollows between the dunes and the excavations dug by the wind were a number of mushrooms:

> White, mobile dunes [without vegetation] are closest to the sea, where the wind brings new sand and often breaks up existing dunes. These change, as vegetation grows closer to the gray dunes [with sand-fixing plants], initially with a thin, low, often unstable cover, which however is eventually able to suppress the tall dune grasses. Many of the dwarf grasses, mosses, lichens and shrubs mentioned may appear in association, and the furthest dunes from the sea in particular will often show an association with Calluna [heather] (Warming 1913, p. 7).

We note in passing that Warming incorporated the terminology adopted by the 1910 Brussels Congress, notably concerning associations.

The second formation is the sandy salt marshes in the north of the island. These are vast expanses covered with water at high tide and clear at low tide. Here, arenicoles and corophiums can be found, a whole range of flora and fauna intermingled with algae, and then a succession of plants immigrate, initially mainly halophytes, which are gradually displaced in several transitional stages, marked in particular by the arrival of ants. Finally, in the clayey salt marshes of the eastern bay and south of Nordby – the island's port city – and in several places on the eastern coast where the water is calmer, particles of clay and organic matter are deposited and fixed by plants. To the south of Nordby, vast mud banks are formed with associations of algae. Where the terrain is higher, saltwort thrives. The plants are home to numerous small snails, which cover the bottom with their excrement. Where the ground rises and becomes drier as one moves away from the shore, more or less marshy meadow plants grow, with clay channels running through them and numerous holes (Warming 1913, p. 8).

Figure 18.2. Salicornia herbacea (S. europaea) *on the island of Fanø (Warming 1906, p. 203)*

A second example is Central and East Jutland. Warming looked at the succession of vegetation since the ice had melted, leading to the development of aspen and birch, then pine forests, supplanted by oak and beech.

On his 50 km journey from Herning to Silkeborg, he observed "the remains of the old oak forests, which have disappeared in the struggle with man and his domestic animals" (Warming 1913, p. 15). In 1844, Silkeborg was home to a modern pulp mill, using hydroelectricity. The natural pines (*Pinus sylvestris*) had disappeared, and the growing ones had been planted 150 years ago (*Pinus montana*). In the Silkeborg region, the sandy hills were covered with beech forests and the wetter hollows with birch and alder. There were also coniferous forests. Himmelbjerget, near Silkeborg, had long been considered the highest peak in Denmark (excluding Greenland and the Faroe Islands), at 147 m[1].

Finally, further south and east, in the Viemose area of southern Sjælland, oak forests were cultivated to provide timber, mainly for shipbuilding. The hazel undergrowth was replaced by beech or elm, maple and ash and a host of shrubs. In season, the undergrowth was carpeted with spring plants, many of which were familiar from temperate forests (anemones, primroses, stellaries, etc.). This environment, originally created for specific needs, was now home to "luxuriant vegetation", wrote Warming (1913, p. 18). Today, conifers tend to be found in the west and deciduous trees in the east.

1 In 2005, Møllehøj was officially recognized as the highest point, at 170.86 m.

Depressions in coastal areas, sometimes in salt meadows (schorres), attracted his attention[2]. Their presence was not without economic consequences.

18.2.4. *Mysterious depressions*

In the salt meadows on clay soil explored by Warming, on the islands or on the western coast of Jutland, irregular, more or less deep, bowl-shaped holes could be found. Some were filled with water, others were dry, and the bottom was covered with dried-out plants. Many were covered with a layer of decomposing algae, Phycochromaceae (blue-green algae grouped with Bacteriaceae in the schizophyte group) and diatoms.

These depressions were therefore at different stages. Warming had no doubt that the sea produced them. However, he said he had long wondered about the formation of those scattered over the meadows at some distance from the beach. He was convinced that they were due to piles of seaweed and eelgrass (kelp) washed ashore by high water. As these plants putrefied, they destroyed the meadow vegetation (several photographs illustrate his point). In several places, where the ground was covered with vigorous turf, it could be seen that nearby plants were affected.

This hypothesis on the formation of these depressions with putrefied algae was set out along with three others by Warming. They were discussed for decades, far beyond Denmark.

Warming's contemporary, Richard Henry Yapp, Chair of Botany at the University of Belfast, studied the marshes of Dovey, an estuary in Wales, and described their plant associations. Yapp considers Warming to have supplied the most important contributions to this question of the origin of observed depressions (Yapp et al. 1917; Yapp 1922). He examines with interest all his hypotheses. They were summarized in 2012 in an American study on marine sedimentology, which cites Warming's pioneering work of 1904. It was possible these depressions formed as a result of uneven colonization by marsh plants. The result was areas without vegetation surrounded by areas of growing vegetation. Or it was possible the depressions were formed on coastlines exposed to "beach ridges", a kind of oblique levee in relation to the shoreline, built up by the swell before the marshes were formed. Perhaps even a case of uneven coastal erosion in small bays later closed by beach ridges. Finally, the fourth hypothesis that Warming eventually adopted was that of direct wave attack on weak points on the marsh surface during storms, resulting in the accumulation of putrefying algae (Bartholdy 2012, p. 157).

---

2 The schorre is upstream of the mudflats, the slikke downstream. It is therefore more often flooded, while the schorre is only covered during high tides.

This hypothesis was also put forward in 1980 in a study of Virginia marshes on the east coast of the United States. Dead wrack forms mats that are carried away by winds, tides and currents. This compacted, smothered vegetation creates depressions devoid of living vegetation. The authors' aim was to study the evolution of these depressions over a longer period (1974–1978) than Warming (Reidenbaugh and Banta 1980).

a) b)

Figure 18.3. *Average growth form of* Spartina alterniflora *(invasive halophilous grass). Between May 20, 1975 (a) and July 23, 1975 (b), mats of rockweed (brown algae) covered the grass (IBIS: Intensive Biometric Intertidal Survey) (Reindenbaugh and Banta 1980, p. 397)*

Warming also suggested that the trampling by cattle could create the weak points that the waves had dug out. This idea was taken up again in 1938 by geographer and geologist Niels Nielsen, who took up the problem of the formation of the Wadden Sea marshes, whether natural or man-made. He pointed out the importance of this question in this coastal part of Southwest Jutland and in the Danish North Sea islands and that the uncertainties were due in particular to the lack of knowledge of the soils (Nielsen 1938, pp. 123–124). Warming emphasized the progress to be made in bacteriology.

18.2.5. *A fierce battle against the sea*

This fundamental research led to economic considerations of the utmost importance for Denmark. This is understandable when depressions of putrefied algae form on pastures or arable land. A more global question is that of the productivity of coastal lands and their conservation, or even extension:

> In the summer of 1889, I visited my native island, Mane [Mandø], and got to know its inhabitants. By observing the formation of the marshes

> and learning about the inhabitants' experiences of them, I realized that there were some interesting biological questions here worth investigating, in particular the contribution of plants to soil formation, the reasons for their belt arrangement and the struggles between different communities. The visit helped to concretize a project I'd had in mind ever since my stay at Lagoa Santa and my visit to Greenland had introduced me to ecological studies, namely to study the plant communities of Denmark (Warming 1906–1919, p. 3).

Between then and the beginning of the 20th century, his knowledge of local plant communities deepened thanks to numerous excursions.

"The harmony between the plant world and surrounding nature is not the subject of as much study as it should be," he laments. Species live in harmony with the environment, in terms of both structure and function. As we have seen, it interprets the concept of epharmony in this sense. Age-old human activities – breeding, cultivation, construction – contribute to this harmony, which it accounts for through a dynamic, historical approach to different plant formations and communities (Warming 1896, p. 376, 1902, p. 391, 1906–1919, p. 4, p. 267, 1909, pp. 369–370).

More generally, this quest for harmony is rooted in Danish culture, art and way of life. The work of neoclassical architect Hansen is marked by the harmony and simplicity of its lines. The quest for harmony is also reflected in everyday life, in a taste for simple things, for family, moral values and identity. Landscape and nature painters give an idealized image of a harmonious, splendid Denmark, with a simple, rural population (Lundbye, Skovgaard). *Hygge*, which comes from a Norwegian word adopted by the Danes in the 18th century, refers to an atmosphere of warmth, comfort and security in the moments of daily life, under the living light (*levende lys*) of candles.

Warming was not a contemplative. He was a scientist who studied the potentialities of environments, discussing the ways in which they could be enhanced through the contributions of science and technology and opened up avenues for progress. For example, he assessed the nutritional value of the marshes of the South West, the value of land protected by dikes and embankments, and the richness of plant and animal communities. He insisted on the need to continue chemical analysis of soils, which was still very incomplete, in order to measure their fertility (Warming 1906–1919, pp. 268–269).

Among the large man-made formations, coastal meadows are remarkable for their high productivity. Fodder plants are grown and exploited, and cattle have been successfully grazed for centuries. Cattle are fattened before being sold. In western Holstein, he describes lush fields of wheat, beans and turnips, plantations of lime, elm, ash, horse chestnut and locust trees. It was a bocage landscape. "It's not surprising, from what has been said, that for centuries the peasants of the marshes have waged a bitter struggle against the sea, on the one hand to force it to produce more of this precious soil, and on the other to prevent what has been formed from being swallowed up". It became a rich, fertile country where poorhouses were rare. In some respects, this Edenic description evokes the pastoral scenes that enchanted Pontoppidan. Warming supported the farmers of Jutland, including financially. This insistence on prosperity built up over centuries by good Danish peasants – by which we mean Danophiles – took on a particular political meaning after the Duchy Wars.

He gave his opinion on the suitability of structures such as dams, banks and embankments, and how they could enhance the fertility of the land by cushioning the effects of the tide and promoting silt deposits. Fertile land was reclaimed from the sea thanks to major works: "Artificial cultivation is practiced throughout the marshes. The first major works have already been undertaken off Ribe and Fanø. In the northwestern parishes of southern Jutland (Hvidding, Rejsby, Brøns), a land reclamation society was set up in 1856". Work stopped for a while. Warming did not mention the Duchy Wars, but all of his Danish readers got the hint. Then, "dry, grassy areas of several hundred acres were reclaimed between the mainland and the islands" (Warming 1906–1919, pp. 271–272). In Holstein (then no longer part of Denmark), fertile land was said to have been reclaimed in this way. It would appear that he did not verify this himself.

The landscape of Jutland was formed when it was developed after the Duchy Wars, in particular as a result of the influx of refugees to the north. The "conquest of the North" involved exploiting territories that had previously been sparsely populated and dominated by moorland (Bazin 1994, p. 130). At the beginning of the 20th century, Warming questioned this dynamic of valorization in coastal areas. He was not in favor of building large dams, which were costly and consumed grazing land. When heavy flooding occurred, they risked being destroyed, wiping out years of land reclamation. Like all Jutland residents on the western coasts, he was well aware of the instability of the coasts and the threat posed by recurrent flooding. There were ecological and economic issues, as well as pedological and floristic considerations. Some indicators are discussed. We could, he suggested, dike when the land is sufficiently high so as not to be flooded during the usual spring floods, or when the clover begins to take hold. In other words, no need to get ahead of yourself. It is better to ask nature itself about the best time to act.

Dams hold back water while dikes prevent it from coming in. Warming recommended sticking to dikes that protect against tides and meteorological events. Traditionally, communities have cooperated to build and maintain them. Since 1851, they have been officially organized as cooperatives. The fact that he suggested that local people should be left to deal with this issue themselves, as was always the case, rather than relying on more technically advanced engineering structures built by engineers, was a source of inspiration for him.

He noted that in the past, dikes were too weak. Now, a network of old and new, inner, outer and intermediate dikes provided several levels of land protection (Warming 1906–1919, p. 273). In Chapter 18 of *Dansk Plantevækst* (1906) on artificial land reclamation, he assessed the advantages and disadvantages of different situations and types of flooding. The aim was to conquer and maintain perennial land:

> Dike construction is an art in itself. In addition to the dike itself, it is often necessary to build locks, gates through which watercourses can flow at low tide, but which close at high tide. When you think of all the changes that man has brought, and brings every year, to the North Sea coast, you understand how true that old saying is: "God made the sea, but Frisia made the shore" ("Deus mare, Friso littora"). To this we may add: "The sea formed the marshes, the wind the dunes" (Warming 1906–1919, p. 275).

Behind the dikes, a succession of vegetation types was observed: "1) the foreshore; 2) the foreland, with its vegetation of dark green sedges (*Carex*) and the native marsh formation; 3) the diked marsh". Trenches divided the fields. Their function to promote drainage of the land and regulate its water retention capacity.

These developments were conducive to the development of new plant communities of ditches, marshes and dikes. Ditches are nutrient-rich environments. Sometimes overgrown with reeds, they hide aquatic and swamp plants that thrive there. Together with several other botanists (notably Vaupell), he listed the plants identified on the west coast of Jutland, specifying their biotope (aquatic, marshy) and sometimes "companion" plants. They were all characteristic of freshwater environments – a good sign for cultivation – with the exception of a few bulrushes, such as sea bulrush, which indicate proximity to the sea.

On the eastern coast of Sylt (one of the Frisian islands), plants such as the marsh zannichellie, whose threadlike leaves and tiny flowers adapted to a submerged life indicate a brackish environment. A perpendicular transect along the coast shows that these brackish-water species are gradually being replaced by freshwater species.

For Warming, the effectiveness of the dikes in desalinating the soil and their artificialization for cultivation and livestock farming was beyond doubt. Years of rainwater leaching and the fact that the sea does not pass over the dike lines have resulted in the formation of perennial grassland, without the need to sow grass seed. These are either grazed meadows, hay meadows or have been converted to arable land. The vegetation is no longer halophilic; it is mesophilic. It is a landscape without beauty, but impressive in its extent, he wrote (Warming 1906–1919, p. 280).

A different type of vegetation developed on the dikes themselves, on clay soil, drier and exposed to the sun, spared by cattle. Today, these dikes are referred to as ecological.

18.2.6. *An active conservationist*

Not surprisingly, Warming was particularly interested in heathland protection. For him, it was not a question of preserving it, in the sense that heathland should not be used or exploited, but of conserving it. Conservation promotes reasoned use and management, with respect for ecological dynamics of which Warming was one of the leading experts. Between 1901 and 1919, several actions and interventions were emblematic of the power his knowledge gave him.

In 1901, he obtained from the new Minister of the Interior, Enevold Sørensen that the great Borris moor in West Jutland should not be forested, but left to rest. "The financial problem was solved by Colonel Emmanuel Prytz of the artillery shooting school, who obtained permission from the Minister of War [Vilhelm Herman Oluf] Madsen to use the moor for a short period in the afternoon for shooting, without affecting the original conservation plan". Hanne wrote of this, "I am so happy for Eugen that his many travels, writings, visits to ministers, members of parliament and others were not in vain".

In 1905, a meeting was held with the Botanical Society, Natural History Society and the Geological Society were invited. Warming invited the great German conservationist Professor Conwentz, from Danzig, to give a lecture on "The conservation of natural monuments". The previous year, he had published a paper on the threats to these "natural monuments" and proposals for their preservation. For Conwentz, these were biological vestiges that preserved the memory of past natural states. But this notion was broadened, insofar as there were no longer any totally untouched landscapes, particularly in Germany. There was a need to come to terms with industrialization and human activity, and invent ways of avoiding excessively negative effects on nature, which was becoming a matter of general interest.

The result of this 1905 meeting was the creation of a "Nature Conservation Committee" charged with protecting rare and unusual animals and plants, as well as other natural features whose existence was threatened. Each of the three associations appointed members, and Warming chaired the committee. In a draft report, he stated the desirability of preserving a larger continuous tract of heather moorland to retain for posterity the proud image of Jutland's poetry, to preserve the memory of the remarkable work of farmers cultivating this country for centuries, to pay tribute to and extend the cultural work of Dalgas and *Hedeselskabet* (Danish Heathland Society, which still exists). "We hope that a large part of Jutland, with its brown heather, oak scrub and moorland, herons, adders and other wildlife, will be declared a protected national park for all" (Conwenth 1904; Prytz 1984, p. 134, pp. 137–138). He may well have been familiar with initiatives in the United States (Yellowstone in 1872, Sequoia and Yosemite in 1890) or even Australia (Royal National Park 1879).

Dalgas co-founded *Hedeselskabet* in 1866 with lawyer and botanist Georg Morville, who acquired land north of Viborg. Dalgas applied his knowledge and practices acquired as a military engineer, to the afforestation of western Jutland. Working on traffic routes, he had a very good knowledge of the soil and the concerns of the local people he consulted. In particular, he had to eliminate the hardpan, a compact, impermeable layer of soil that forms a sort of crust, preventing root growth and water circulation in the soil. The solution is to marl the soil, making it more calcareous. This makes them less acidic and more suitable for farming.

The aim of the Danish Heathland Society was to enhance the value of this plant formation by irrigating and planting trees and hedging. The society also exploited peat bogs. Subsidized by the State, it created pine plantations and provided plants free of charge and bought land to establish its own plantations[3]. Depending on the local situation, the analyses suggested conserving the heathland, reforesting it, modifying the soil properties, irrigating or draining it. As we all know, when it comes to ecology the best option is never written in advance. Several solutions sometimes need to be combined, after consultation with users, which is what the Danish society did.

In 1917, as part of the first Nature Conservation Act passed by Parliament, a five-member "Nature Conservation Council" was created, three of whom were elected by the Ministry of Education after consultation with the Faculty of Mathematics and Natural Sciences. Warming was elected president.

Finally, in August 1919, he replied to the Ministry of Agriculture, as conservation issues fell within its remit: "On June 19, the High Ministry sent me a request for advice on the future operation of the Jaegersborg deer park. It concerns the town of Dyrehaven, which I've known for a long time and which is located in

3 In 1916, it had 9,000 members and 47 paid civil servants. It had helped to plant 80,000 ha (6,500 ha of its own). A total of 70,000 ha of water-saturated land were drained.

the extreme south of the country". Today, the former Royal Hunting Reserve Dyrehaven, north of Copenhagen, is a forest park that is home to Cervidae.

Therefore, the powers that be sought the advice of a much honored scholar, with universally recognized scientific and institutional stature. As he grew older, Warming became more involved in Danish nature conservation issues with a view to enhancing the economic, social, cultural and ecological value of the region.

18.3. "The white man dressed all in black"

Throughout his life, Warming never hesitated to face up to the dangers and sometimes difficult conditions of long expeditions, in order to make his contribution to the colonial ecological project of the Danish Crown. After his period of travel, he used his institutional position and scientific authority to support a new generation of traveling naturalists.

In the foreword to *Botany of Iceland* (1912), signed by Kolderup Rosenvinge and Warming, it is stated: "It was mentioned in the preface to *Botany of the Faroes* [1901] that, to complete the work, Iceland should be an island among the dependencies of the Danish kingdom in the Atlantic which was most in need of thorough and systematic research as far as botany was concerned". An appeal was made to the younger generation of botanists. The task ahead was enormous and would take several years. It ended in 1949 with the publication of a flora of the Reykjanes peninsula by the phytocenologist and ecologist Hadač (1949). Warming's contributions were included up to 1945.

At the beginning of the 20th century, Børgesen, who had long been involved in the study of Faroese botany, dealt with cultivated plants on whom Warming had long been able to rely, discusses cultivated plants with a view to the economic development of the archipelago, in particular through the planting of woody species and the cultivation of potatoes, which remained low-yielding (Gardening and tree-planting, 1908). Agricultural issues were also addressed by another contributor, Feilberg (Some notes on the agriculture of the Færöes, 1908).in this archipelago where two-thirds of resources come from fishing (Warming 1901–1903–1908, p. 1027 *ff*, p. 1044 *ff*).

This process of appropriating the environment and resources of conquered territories for the benefit of Europe is part of an age-old enterprise whose consequences are multiple and complex. According to environmental historian Richard Grove, the conquest of the tropics gave rise to Western ecology, both in the scientific field and in that of environmental awareness, as studied by Mathis for the industrial era in England (Grove 1996, 2013; Mathis 2010).

Arctic exploration can be seen as playing an equally important role in the birth of ecology in Denmark. For almost three centuries, Greenland's colonial history has been strongly linked to the Kingdom of Denmark. The Carlsberg Foundation's support for the marine biology of the West Indies and numerous other projects was all part of this colonial logic. The effects on the development of Danish ecology, of which Warming was a key player, are undeniable.

It was also part of a centuries-old history of missionary work.

In Greenland, "the white man dressed all in black, the black Lutheran colonialist man" refers to Pastor Hans Egede, "the apostle of Greenland", in the early 18th century (Leine 2013; Pedersen 2019, p. 60). An evangelist, he was sent on missions with the approval of King Frederick IV, who intended to re-establish Danish sovereignty in Greenland.

Warming was not a clergyman like his father, but at no time did he cast a critical eye on the enterprise of the Danish colonial empire, which he even ardently supported. A religious vision of the world, shared by many Danish naturalists, was consistent with his scientific work.

Conclusion to Part 5

Warming developed a dynamic approach to plant communities modified by human activities, which is a form of ecological engineering.

"He was especially concerned with the acute economic problems involved in agricultural development in Jutland, a solution to which involved the use of plants in controlling moving sand and the conversion of heath into tillable land" (Coleman 1986, p. 186). One of his students, August Mentz, wrote a book in 1909 on nature conservation and defended his thesis in 1912 on the vegetation of the peat bogs and marshes of Jutland (Mentz 1909, 1912). He became one of the pioneers of nature conservation, focused on heathland restoration. While Warming's work and actions had an ecological dimension in terms of his work and actions, he was, like his student, a conservationist and utilitarian. While he deplored certain actions in the field of forest and land management, for him, the overall assessment of the impact of human activities on the environment was far from negative. It even contributed to the harmony that reigns in nature.

Recognized by his contemporaries as the founder of ecology, he was a scientific reference and one of the best promoters of the Danish colonial ecological project. A cohort of students and disciples followed in his footsteps. In 1915, the painter K.E. Larsen, who was often called upon by public and private institutions, painted Warming's portrait for Carlsberg's biology department, with the text: "An outstanding systematist and a pioneer in the field of ecology".

Warming had an authoritarian and paternalistic conception of his responsibilities. He kept a close eye on the affairs and choices of the institutions for which he was responsible or within which he had a certain amount of power, whether at the Carlsberg Foundation, at the university or in the scientific academies of which he was a member. A man of conviction, he had strong sympathies or antipathies, which he expressed forthrightly, and he did not hesitate to stir up controversy or take part in it. Fox Maule, from the University of Copenhagen, wrote in the *Dansk Biografisk Leksikon* (Danish Biographical Lexicon):

> He was a man of strong sympathies and antipathies, who could be harsh in his judgment of those he disliked, and cutting in his discourses during the fairly numerous controversies in which he took part. On the other hand, he was an influential supporter of those in whom he detected ability and drive (Fox Maule 2011b).

Although he did not commit himself publicly to politics, he made no secret of his Christian and conservative ideas. So the progressive current of Brandesianism did not sit well with him. "Brandes deplored the pervasiveness of religion, believing that the anticlerical struggle needed fervent apostles in the service of free thought" (Largueche 2015, Chapter VI, p. 148). Warming saw the intellectual Georg Brandes as a dangerous liberal radical, a positivist propagator of an atheism that risked spreading among Copenhagen academics and turning away readers of the conservative newspaper *Dagbladet*, who may be tempted by socialist ideas.

For many years, Warming maintained close relations with German colleagues, but became more closely associated with the English and French in his later years. Patriotic and anti-Prussian, the attachment of Northern Schleswig to Denmark, following plebiscites in accordance with the terms of the Treaty of Versailles (June 28, 1919), would have been one of his last joys. In 1920, despite his advanced age, he was keen to take part in the excursion to Sonderjylland (southern Jutland) planned by the Botanical Society, a region that was dear to his heart, and which had already become part of his homeland.

His wife, who had always supported her husband and shared his religious faith, died of heart disease on October 31, 1922, less than a year after a grand celebration of their golden wedding anniversary on November 10, 1921. It was attended by almost 60 people at the shooting range, 40 years after the party organized in the same place for his departure to Stockholm. Numerous letters and gifts arrived from Denmark, Europe and the United States, taking him back in time: one from Ribe with a letter from his comrade Julius Pedersen. Warming thanked him and remembered his mother, who welcomed him at her table when he was a schoolboy. Brazil manifested itself through a letter from Frederik Riise, commissioner for Denmark at the Universal Exhibition in Rio de Janeiro (September 7, 1922 to March 23, 1923). He informed him that he would be making excursions to the Lagoa Santa region and visiting Lund's grave (Riise 1924; Prytz 1984, p. 168).

In the winter of 1921–1922, Warming fell seriously ill, recovered and resumed his activities. Admitted to Rigshospitalet in March 1924 – set up in 1910 at its present location – he continued to write letters from his hospital room. He died on

April 2, 1924, two days after an operation. The entire Danish scientific community attended his funeral, and a bronze statue was erected in front of the botany laboratory in 1927. A commemorative plaque was unveiled in Lagoa Santa in 1933, and another in 1950 at the former Mandø presbytery. Despite the war, the 100th anniversary of his birth was celebrated in 1941 on the initiative of his granddaughter Signe Prytz. On November 3, 1966, the event was celebrated again. “His name may not mean much to many journal readers today, but botanists at home and abroad remember him” (Prytz 1984, p. 181).

Figure C5.1. *Commemorative plaque for Warming in Mandø (Erik Christensen, CC BY-SA 3.0, public domain via Wikimedia Commons)*

General Conclusion

Eugenius Warming's inner strength, work capacity and pugnacity were exceptional. He must have experienced moments of discouragement, perhaps loneliness and despondency, after the news of his mother's death, when he had not yet unpacked his luggage in Lagoa Santa. He could have given up and returned to Denmark on the first boat. Not only did he stay, but he extended his stay by several months beyond what was originally planned.

He was no adventurer. From his first sojourn in Brazil to his life at university in Copenhagen, this traveler never lost his way, he always knew where he was going, he mapped out his path and built his career, as a *sagax et astutus* man. Essayist Paul-Dubois wrote in *Revue des deux mondes*:

> Skilful, circumspect, willingly opportunistic, the *Danus astutus* of the old authors combines good sense with first-rate practicality, but in everything he displays more tenacity than temerity, more industry than ambition; his character leads him to the defensive rather than the offensive; he is by nature conservative (Paul-Dubois 1909, p. 660).

Eugenius had no shortage of ambition. His intellectual and institutional career attests to this. Opportunistic, no doubt, but in the sense that he knew how to take advantage of the circumstances and opportunities available to him, such as his trip to Brazil or his position at Stockholm University. Tenacious but cautious, an authentic conservative patriot – by nature or family tradition – he was the exact opposite of Didrichsen, a social democrat portrayed by his peers as whimsical and inconsistent. Didrichsen was placed in Warming's path to a position at the University of Copenhagen. It was just a detour. It became clear that Eugenius was as loyal in friendship as he was in enmity, as evidenced by his letters and the echo given by scientific publications and the press.

He was more offensive than defensive, skillful and circumspect, including in the many controversies and polemics in which he was not known for dodging, when he judged them to be worthwhile: establishing his authority over the administration of the botanical garden, forging useful alliances, vigorously defending convictions and conceptions, and so on.

In Lagoa Santa, the young, inexperienced student and orphan was already thinking about his career and making his work known. The publication of the first elements of his flora of Central Brazil, when he was only 26, was hailed by his peers in Denmark and Europe. Years later, as a prominent Danishman settled in Copenhagen with his family, a man of the field and of networks, a relentless worker driven by a passion for passing on his knowledge, he won international recognition with his treatise on plant ecology. In Brazil, he went down in history as the author of the first study of tropical ecology. His work on Greenland was pioneering.

He was a man of the 19th century through his scientific training and his culture, rooted in Scandinavianism. His Lutheran morality paints an austere picture of a man of science known for his probity. His work of compilation and synthesis was accompanied by a process of conceptualization and clarification of a botanical geography inspired by Humboldt, from which he drew the lines of an innovative ecological program.

He was also a man of the 20th century, in tune with the ecology driven by a new generation in the United States, whose pioneers he inspired. He knew how to call on new collaborators and identify the authors who mattered for the nascent ecology. His contributions saved precious time for the founders of animal ecology in the 1920s and 1930s.

He understood that ecology had to integrate the methods of experimental physiology, the contributions of genetics and the concepts of Darwin's evolutionary theory. Although he took them into account at a theoretical and reflexive level, he only succeeded imperfectly in incorporating them into his ecological paradigm, resulting in a lack of clarity in the 1909 English version. *Oecology of Plants* gives the impression of a kind of stratification. Beneath the recent layer of ecology of the 20th century lie the thick strata of ecological botanical geography of the 19th century.

However, it was the very essence of ecology that he founded, a "self-conscious" ecology. The term self-conscious was used in the 1940s by US zoologists in a study of animal ecology to identify this historic moment (Allee et al. 1949). Indeed, committed to solving the problem of the multiplication of systems and the difficulty of finding synonymies that would win consensus, he did not proceed by accumulation as others had done before him. He conceived a level of integration, that of “association groups”, themselves divided into a number of major ecological series (hydrophytes, xerophytes, halophytes, mesophytes). Concerning the question of the geographical distribution of plants and their grouping into communities, “Warming’s answers gave doctrinal coherence to and helped delineate the fundamental problems of ecology” (Coleman 1986, p. 186). He thus created the conditions for thinking about the concept of the ecosystem.

To this day, this process of creating ever more integrated units is a success story, and, at the same time, the subject of bitter debate. Other proposals include the ecocomplex by ecology professor Patrick Blandin. The ecocomplex represents a system of interdependent ecosystems in a territory whose structures, dynamics and functionalities are the result of interwoven natural and human histories: “A space bearing a more or less anthropized set of interactive ecosystems” (Blandin 1992, p. 274; Matagne 2014). This has important theoretical and practical implications, particularly for ecological engineering, which works on complex systems.

After the publication in 1909 of *Oecology of Plants*, which had been delayed several times, Professor Warming seemed freed from an important task. He was finally able to resume his long-abandoned studies, some of which he had begun during his thesis, and to devote himself more intensively to his work in botany and ecological botanical geography for the Danish Crown.

In his twilight years, he remobilized his early studies on plant life-forms and ontogenesis. He gathered around him a new generation of Danish scientists and students, some of whom became his followers; he had the desire to become a school of thought. He succeeded, far beyond Denmark, in building “the paradigm of ecological study” (Goodland 1975, p. 244), which served as a framework for the work of the first specialists in ecology, European and American ecologists. “The Danish school has been a pioneer in ecology and has developed its ideas with a marked independence”, which drew attention to the value of knowing Danish vegetation, wrote William Gardner Smith in 1914, brother and continuator of the work of Robert Smith, founder of ecology in Great Britain (Smith 1914, p. 65).

In the introduction to this book, we raised the question of the unpredictable, the unexpected and the uncertain, which could have marked out Warming’s career. He seemed to have exercised a great deal of control over his life and career, not as if his destiny were sealed, but because he himself wrote the most important pages.

A child born on a tiny, windswept island was lucky enough to be born into a privileged economic and cultural environment, who nonetheless had to face unexpected and unforeseeable trials. Some were dramatic, at a family level, with the loss of his mother while he was far away, of several children and of his partner. His first big trip was certainly unexpected, but his early naturalist skills made him the best candidate. He then worked hard to make his other trips possible, seizing opportunities, getting the right support and using his own influence. His appointment as professor at the University of Copenhagen was uncertain, but his voluntary exile in Stockholm enabled him to exercise his talents as an administrator. Back in Copenhagen, he soon became a key figure. Occupying all the key positions, Professor Warming also built careers, for which many of his students were indebted to him.

As for his extraordinary professional success and the fame he enjoyed during his lifetime, he always seemed surprised by it, and never sought it out. He declared that his only aim was to work for his fellow countrymen and for the greatness of Denmark.

References

Acot, P. (1983). Darwin et l'écologie. *Revue d'histoire des sciences*, 36(1), 33–48.

Acot, P. (1988). *Histoire de l'écologie*. PUF, Paris.

Acot, P. (1997). The Lamarckian cradle of scientific ecology. *Acta Biotheoretica*, 45(3–4), 185–193.

Acot, P. (1998). Botanical geography. In *The European Origins of Scientific Ecology*. Gordon and Breach Publishers, Amsterdam.

Acot, P. and Drouin, J.-M. (1997). L'introduction en France des idées de l'écologie scientifique américaine dans l'entre-deux-guerres. *Revue d'histoire des sciences*, 50(4), 461–480.

Agassiz, L. (1869). *Voyage au Brésil*. Librairie Hachette et Cie, Paris.

Alexandre, F. and Génin, A. (2011). *Géographie de la végétation terrestre. Modèles hérités, perspectives, concepts et méthodes*. Armand Colin, Paris.

Allain, Y.-M. (2000). *Voyages et survie des plantes au temps de la voile*. Champflour, Marly-le-Roi.

Allee, W.C., Emerson, A.E., Park, T., Park, O., Schmidt, K.P. (1949). *Principles of Animal Ecology*. W.B. Saunders Company, Philadelphia.

André, M.-F. (2000). Chronique arctique et antarctique. *Norois*, 186, 235–268.

Anker, P. (2002). *Imperial Ecology: Environmental Order in the British Empire, 1895–1945*. Harvard University Press, Cambridge.

Anker, P. (2011). Plant community, Plantesamfund. In *Ecology Revisited, Reflecting on Concepts, Advancing Science*, Schwarz, A. and Jax, K. (eds). Springer, New York.

Appell, G. (1948). L'éducation au Danemark. *Enfance*, 1(3), 275–281.

Aybar Camacho, C.L. and Lavado Casimiro, W. (2017). Atlas de zonas de vida del Peru, Guía explicativa. Report, Servicio Nacional de Meteorología e Hidrología del Perú (SENAMHI), Dirección de Hidrología, Lima.

Aymonin, G.G. and Keraudren-Aymonin, M. (1990). Autour de Gaston Bonnier et de son œuvre : essai documentaire comparatif. *Bulletin de la Société Botanique de France. Lettres Botaniques*, 137(2–3), 125–138.

Baillon, H. (1888). *Histoire des plantes. Monographie des Droséracées, Tamaricacées, Salicacées, Batidacées, Podostémacées, Plantaginacées, Solanacées, Scrofulariacées.* Librairie Hachette et Cie, Paris.

Balslev, V. (1900). *Praktisk Haelpebog for begyndende Naturhistorielaerer*, 3rd edition. Lehmann & Stages Forlag, Copenhagen.

Bange, C. (2004). Le botaniste Alexis Jordan (1814–1897) à la Société linnéenne de Lyon. *Bulletin de la Société linnéenne de Lyon*, 73(1), 7–24.

Bartholdy, J. (2012). Salt marsh sedimentation. In *Principles of Tidal Sedimentology*, Davis, R.A. Jr, and Dalrymple, R.W. (eds). Springer, New York.

Bazin, P. (1994). Les brise-vents au Danemark. *Revue forestières française*, 46, 130–138.

Beauvais, J.-F. and Matagne, P. (1999). Le concept de zone de vie de Holdridge : un point de vue tropical en écologie. *Revue internationale de la Société Française d'Écologie*, 29(4), 557–564.

Beauvais, J.-F. and Matagne, P. (2014). *Une écologie du Nouveau Monde. Les Zones de vie d'Holdridge*. Édilivre, Saint-Denis.

Beneden, P.-J. (1875). *Les commensaux et les parasités dans le règne animal*. Hayer, Bruxelles.

Benest, G. (1990). Le Laboratoire de Biologie végétale de Fontainebleau : création et héritage de Gaston Bonnier. *Bulletin de la Société Botanique de France. Lettres Botaniques*, 137(2–3), 115–119.

Binet, P. (1978). Introduction : caractéristiques physiologiques liées à l'halophilie et à la résistance aux sels. *Bulletin de la Société Botanique de France. Actualités Botaniques*, 125(3–4), 73–93.

Blandin, P. (1992). De l'écosystème à l'écocomplexe. Entre nature et société, les passeurs de frontière. In *Sciences de la nature, sciences de la société*, Jolivet, M. (ed.). CNRS Éditions/Les passeurs de frontières, Paris.

Blaringhem, L. (1905). La notion d'espèce et la théorie de la mutation. *L'année psychologique*, 12, 95–112.

Blixen, K. (1961). *Le dîner de Babette et autres contes*. Gallimard, Paris.

Boergesen, F. and Paulsen, O. (1900). *La végétation des Antilles Danoises*. Paul Dupont Éditeur, Paris.

Bonnier, G. (1890). Cultures expérimentales dans les Alpes et les Pyrénées. *Revue générale de botanique*, 2, 513–546.

Bonnier, G. (1895a). Les plantes de la région alpine et leurs rapports avec le climat. *Annales de géographie*, 4(17), 393–413.

Bonnier, G. (1895b). Influence de la lumière électrique continue sur la forme et la structure des plantes. *Revue générale de botanique*, 7, 241–257.

Bonnier, G. (1913). Chronique scientifique. Les Alpes. In *La Revue hebdomadaire et son supplément illustré paraissant le samedi*. Librairie Plon, Paris.

Bonnier, G. and Flahault, C. (1878). Observations sur les modifications des végétaux suivant les conditions physiques du milieu. *Annales des sciences naturelles Botanique*, 6(7), 93–101, 113–118, 124–125.

Børgesen, F. and Paulsen, O. (1898). *Om Vegetationen paa De dansk-vestindiske Øer*. Nordisk Forlag, Copenhagen.

Braun-Blanquet, J. (1929–1930). L'origine et le développement des Flores dans le massif central de France avec un aperçu sur les migrations des Flores dans l'Europe sud-occidentale. *Société linnéenne de Lyon, Société botanique de Lyon et Société d'anthropologie et de biologie de Lyon réunies*, 75, 1–73.

Braun-Blanquet, J. and Furrer, E. (1913). Remarques sur l'étude des groupements de plantes. *Bulletin de la Société languedocienne de géographie*, 36, 20–41.

Britton, N.L. and Brown, A. (1913). *An Illustrated Flora of the Northern United States, Canada and the British Possessions*. Charles Scribner's Sons, New York.

Brockmann-Jerosch, M.C. (1903). *Geschichte und Herkunft der schweizerischen Alpenflora. Eine Übersicht über den gegenwärtigen Stand der Frage*. Wilhelm Engelmann, Leipzig.

Brockmann-Jerosch, H. and Rubel, E. (1912). *Die Einteilung der Pflanzengesellschaften nach ökologisch-physiognomischen Gesichtspunkten*. Wilhelm Eugelmann, Leipzig.

Brongniart, A. (1844). Examen de quelques cas de monstruosité végétale propres à éclaircir la structure du pistil et l'origine de l'ovule. *Annales des sciences naturelles Botanique*, 3(2), 20–32.

Bronwen, D. (2009). L'idée de "race" et l'expérience sur le terrain au XIXe siècle : science, action indigène et vacillations d'un naturaliste français en Océanie. *Revue d'histoire des sciences humaines*, 21(2), 175–209.

Bruyant, C. (1898). Introduction à la faune de l'Auvergne. Notes de géographie biologique. *Revue scientifique du Bourbonnais et du Centre de la France*, 5–17.

Bruyant, C. (1899). Prodrome d'une géographie zoologique du Plateau Central. *Bulletin historique et scientifique de l'Auvergne*, 119–125.

Buchmann, N. (2002). A.F.W. Schimper: From the "ecology of plant distribution" to the "functional ecology of terrestrial ecosystems". *Endeavour*, 26(1), 3–4.

Buican, D. and Grimoult, C. (2011). *L'évolution. Histoire et controverses*. CNRS Éditions, Paris.

Butters, F.K. (1917). Contribution from the Gray herbarium of Harvard University. Taxonomic and geographic studies in North American ferns. *Rhodora, Journal of the New England Botany Club*, 19(225), 169–216.

de Candolle, A.-P. (1820). *Essai élémentaire de Géographie Botanique*. Dictionnaire des sciences naturelles/Levrault, Paris/Strasbourg.

de Candolle, A. (1855). *Géographie botanique raisonnée ou exposition des faits principaux et des lois concernant la distribution géographique des plantes de l'époque actuelle*. Librairie de Victor Masson/Librairie Allemande de J. Kessmann, Paris/Geneva.

de Candolle, A. (1874). Constitution dans le règne végétal des groupes physiologiques applicables à la géographie botanique ancienne et moderne. *Archives des sciences physiques et naturelles*, 50, 5–42.

Capus, G. and de Rochebrune, A.-T. (1879). *Le guide du naturaliste herborisant, conseils sur la récolte de plantes, la préparation des herbiers et les herborisations aux environs de Paris, dans les Ardennes, la Bourgogne*, 2nd edition. J.-B. Baillière, Paris.

Capus, G. and de Rochebrune, A.-T. (1883). *Guide du naturaliste préparateur et du voyageur scientifique*. J.-B. Baillière, Paris.

Carlsen, J., Olrik, H., Starcke, C.U. (1900). *Le Danemark, État actuel de sa civilisation et de son organisation sociale*. Det nordisk Forlag, Ernst Bojesen, Copenhagen.

Carré, F. (1976). Les îles orientales et septentrionales de la Frise. *Norois*, 90, 253–262.

Cassanello, A. (2008). La Commission scientifique du Nord et les relations de voyage de Xavier Marmier et de Léonie d'Aunet. In *Le(s) Nord(s) imaginaire(s)*, Chartier, C. (ed.). Imaginaire-Nord, Montreal.

Cavalcanti, B.M. (2019). A cafeicultura e a Estrada União e Indústria: Transformações espaciais e desenvolvimento desigual em meados do século XIX. *Terra Brasilis* [Online]. Available at: http://journals.openedition.org/terrabrasilis/3852 [Accessed 18 November 2022].

Cavassan, O. and Weiser, L. (2020). Eugen warming: Um dinamarquês desvenda o cerrado brasileiro. *Filosofia e História da Biologia*, 15(2), 179–193.

Chalmer, A.F. (1987). *Qu'est-ce que la science ? Popper, Kuhn, Lakatos, Feyerabend*. La Découverte, Paris.

Christensen, C. (1924–1926). *Den danske Botaniks Historie med Tilhørende Bibliografi*. H. Hagerups Forlag, Copenhagen.

Christensen, C. and Hansen, A. (1913–1963). *Den danske botaniske Litteratur 1880–1911, 1912–1939, 1940-5*. H. Hagerup/Ejnar Munksgaard, Copenhagen.

Cittadino, E. (1990). *Nature as laboratory, Darwinian Plant Ecology in the German Empire (1880–1900)*. Cambridge University Press, Cambridge.

Clements, F.E. (1905). *Research Methods in Ecology*. The University Publishing Company, Lincoln.

Clements, F.E. (1916). *Plant Succession: Analysis of the Development of Vegetation*. Carnegie Institution of Washington, Washington.

Cohen, C. (2004). Balzac et l'invention du concept de milieu. In *Balzac géographe, territoires*, Dufour, P. and Mozet, N. (eds). Christian Pirot Éditeurs, Saint-Cyr-sur-Loire.

Coleman, W. (1986). Evolution into ecology? The strategy of warming's ecological plant geography. *Journal of the History of Biology*, 19(2), 181–196.

Collet, S. (2019). L'évolution des espèces sociales dans *La Comédie humaine* de Balzac. *Arts et Savoirs* [Online]. Available at: http://journals.openedition.org/aes/2317 [Accessed 14 February 2022].

Collins, J.P., Beatty, J., Maienschen, J. (1986). Introduction: Between ecology and evolutionary biology. *Journal of the History of Biology*, 19(2), 169–180.

Conry, Y. (1974). *L'introduction du darwinisme en France au XIXe siècle*. Vrin, Paris.

Conwenth, H. (1904). *Die Gefährdung der Naturdenkmäler und Vorschläge zu ihrer Erhaltung. Denkschrift, dem Herrn Minister der geistlichen, Unterrichts und Medizinal Angelegenheiten überreicht*. Gebrüder Borntraeger, Berlin.

Corsi, P. (1988). *The Age of Lamarck, Evolutionary Theory in France 1790–1830*. University of California Press, Berkeley.

Cosson, E. (1872). *Instructions sur les observations et les collections botaniques à faire dans les voyages*. Imprimerie de E. Martinet, Paris.

Costa, F.A.P.L. (2005). Ecologia de comunidades. *La Insignia* [Online]. Available at: https://www.lainsignia.org/2005/febrero/ecol_006.htm [Accessed 12 August 2022].

Cowles, H.C. (1899). *The Ecological Relations of the Vegetation on the Sand Dunes of Lake Michigan*. University of Chicago Press, Chicago.

Cowles, H.C. (1909). Book reviews, ecology of plants. *Botanical Gazette*, 48(2), 149–152.

Cowles, H.C. (1911). The causes of the vegetational cycles. *Annals of the Association of American Geographers*, 1, 3–20.

Cuvier, G. (1830). *Les révolutions de la surface du globe et sur les changements qu'elles ont produits dans le règne animal*, 6th edition. Edmond d'Ocagne, Paris, Amsterdam.

Da Gloria, P., Neves, W.A., Hube, M. (eds) (2017). *Archaeological and Paleontological Research in Lagoa Santa. The Quest for the First Americans*. Springer International Publishing, New York.

Daniëls, F.J.A. (1982). *Vegetation of the Angmagssalik District, Southeast Greenland, IV. Shrub, Dwarf Shrub and Terricolous Lichens*. Commission for Scientific Research in Greenland. Copenhagen.

Darwin, C.R. (1859). *On the Origin of Species by Means of Natural Selection, or Preservation of Favoured Races in the Struggle for Life*. John Murray, London.

Darwin, C. (1872). *Om Arternes Oprindelse ved Kvalitetsvalg eller ved de heldigst stillede Formers Sejr i Kampen for Tilværelsen*. J.P. Jacobsen, Copenhagen.

Darwin, C. (1875). *Insectivorous Plants*. John Murray, London.

Darwin, C. (1881). *The Formation of Vegetable Mould through the Action of Worms*. John Murray, London.

Darwin, C. and Wallace, A.R. (1858). On the tendency of species to form varieties; and on the perpetuation of varieties and species by natural means of selection, communicated by Lyell C. and Hooker J.D. *Zoological Journal of the Linnean Society*, 3, 45–62.

Davanne, A. (1867). *Exposition universelle de 1867. Rapport du jury international. Épreuves et appareils de photographie*. Imprimerie P. Dupont, Paris.

Davy de Virville, A. (ed.) (1954). *Histoire de la botanique en France*. Société d'édition d'enseignement supérieur, Paris.

Debourdeau, A. (2016). Aux origines de la pensée écologique : Ernst Haeckel, du naturalisme à la philosophie de l'Oikos. *Revue Française d'Histoire des Idées Politiques*, 44, 33–62.

Decaisne, J. (1858–1862–1875). *Le jardin fruitier du Muséum, ou Iconographie de toutes les espèces et variétés d'arbres fruitiers cultivés dans cet établissement*. Firmin Didot frères, Paris.

Delavigne, A.E. (2003). La viande verte : un écologisme à la danoise. In *Bienfaisante nature*, Dubost, F. and Lizet, B. (eds). Le Seuil, Paris.

Deléage, J.-P. (1991). *Une histoire de l'écologie, une science de l'homme et de la nature*. La Découverte, Paris.

Denchev, T.T., Knudsen, H., Denchev, C.M. (2020). The smut fungi of greenland. *MycoKeys*, 64 [Online]. Available at: https://doi.org/10.3897/mycokeys.64.47380 [Accessed 12 October 2021].

De Wildeman, É. (1910). *Actes du IIIe congrès international de botanique de Bruxelles*. De Boeck, Brussels.

Domeier, K. and Haack, M. (1963). *Schleswig-Holstein 1652. Die Landkarten von Johannes Mejer, Husum, aus der neuen Landesbeschreibung der zwei Herzogtümer Schleswig und Holstein von Kaspar Danckwerth 1652. Mit einer Einleitung von Christian Degn*. Heinvetter, Hamburg-Bergedorf.

Dos Santo, N.C. (1923). *O naturalista: Biografia do Dr. Pedro Guilherme Lund*. Impresa Oficial, Belo Horizonte.

Dreyfus-Brisac, E.-P. (1878). *Société pour l'étude des questions d'enseignement supérieur*. Université de Bonn, Paris.

Drouin, J.-M. (1991). *Réinventer la nature. L'écologie et son histoire*, 2nd edition. Flammarion, Paris.

Drouin, J.-M. (1993). Une espèce de livre vivant : le rôle des jardins botaniques d'après Augustin-Pyramus de Candolle. *Saussurea*, 24, 37–46.

Drouin, J.-M. (1994). Histoire et écologie végétale : les origines du concept de succession. *Écologie*, 25(3), 147–155.

Drude, O. (1884). *Die Florenreiche der Erde: Darstellung der gegenwärtigen Verbreitungsverhältnisse der Pflanzen; Ein Beitrag zur vergleichenden Erdkunde.* Ergänzungsheft, Petermanns Mitteilungen, Justus Perthes, Gotha.

Drude, O. (1886–1887). *Atlas der Pflanzenverbreitung.* Berghaus, Physikalischer Atlas, Gotha.

Drude, O. (1897). *Manuel de géographie botanique*, 1st edition. Klincksieck, Paris.

Drude, O. (1913). *Die Ökologie der Pflanzen.* F. Vieweg, Brunswick.

Drude, O. (1914). Die Stellung der physiognomischen Ökologie. *Bericht über die Zusammenkunft der Freien Vereinigung für Pflanzengeographie und Systematische Botanik*, 11, 9–13.

Dübeck, I. (1997). État et Églises au Danemark. In *État et Églises dans l'Union européenne*, Robbers, G. (ed.). Nomos Verlag, Baden-Baden.

Dufour, L. (1914). Le laboratoire de biologie végétale de Fontainebleau. *Revue générale de botanique. Travaux de biologie végétale. Livre dédié à Gaston Bonnier par ses élèves et amis*, T. 25 bis, 1–9.

Dupuy, M. (1997). La diffusion de l'écologie forestière en France et en Allemagne, 1880–1980. PhD Thesis, Université Michel de Montaigne-Bordeaux 3, Bordeaux.

Durand, F. (1957). La figure du Viking dans la littérature scandinave. *Annales de Normandie*, 7(1), 3–33.

Dureau de la Malle, A. (1825). Mémoire sur l'alternance ou sur ce problème : la succession alternative dans la reproduction des espèces végétales vivant en société, est-elle une loi générale de la nature. *Annales des sciences naturelles*, 5, 353–381.

Dyball, R. and Carlsson, L. (2017). Ellen swallow Richards: Mother of human ecology? *Human Ecology Review*, 23(2), 17–28.

Ebach, M.C. (2022). Les origines de la biogéographie : un point de vue personnel. In *La biogéographie une approche intégrative de l'évolution du vivant*, Guilbert, É. (ed.). ISTE Editions, London.

Egerton, F.N. (2018). History of ecological sciences, Part 61B: Terrestrial biogeography and paleobiogeography, 1840–1940s. *The Bulletin of the Ecological Society of America*, 100(1), e01465 [Online]. Available at: https://doi.org/10.1002/bes2.1465 [Accessed 14 February 2022].

Elizondo, S. (1990). Homogénéité morphologique de la population de Lagoa Santa. *Bulletins et Mémoires de la Société d'Anthropologie de Paris*, 2(2), 113–116.

Engelstoft, P. and Dahl, S. (eds) (1943). Grundlagt af C.F. Bricka redigeret af Povl Angelstoft under Medvirkning af Svned Dahl, Udgivet med støtte af Carlsbergfondet. *Dansk Biografisk Lexikon*, 25.

Engler, A. (1902). *Über die neueren Fortschritte der Pflanzengeographie, verzeichnis der besprochenen Schriften. Botanische Jahrbücher fur Systematik Pflanzengeschichte und Pflanzengeographie*. Verlag von Wilhelm Engelmann. Liepzig.

Feldmann, J. (1938). Recherches sur la végétation marine de la Méditerranée, la côte des Albères. *Revue algologique*, 10(1–4), 1–340.

Ferngren, G.B. (2013). *Science and Religion: A Historical Introduction*. John Hopkins University Press, Baltimore, London.

Ferri, E. (1973). *Lagoa Santa e a vegetação de cerrados brasileiros*. Livraria Itatiaia Editora Ltda/Editora da Universidade de São Paulo, Belo Horizonte.

Fischer, J.-L. (ed.) (1999). *Le jardin entre science et représentation*. Comité des Travaux Historiques et Scientifiques, Paris.

Flahault, C. (1900). *Actes du 1e Congrès International de Botanique tenu à Paris à l'occasion de l'Exposition Universelle de 1900*. E. Perrot, Lons-le-Saunier.

Fogh, K.T. (1915). *Hejmdal. Gennem et halvt aarhundrede*. Udgivet, af hejmdal og ripensersamfundet, Copenhagen.

Fol, J.-F. (1978). *Les pays nordiques aux XIXe et XXe siècles*. PUF, Paris.

Foreningsmøder (1882). *Meddelelser fra den Botaniske forening i Kjøbenhavn*. H. Hagerups Boghandel, Copenhagen.

Foreningsmøder (1884). *Meddelelser fra den Botaniske forening i Kjøbenhavn*. H. Hagerups Boghandel, Copenhagen.

Foreningsmøder (1886). *Meddelelser fra den Botaniske forening i Kjøbenhavn*. H. Hagerups Boghandel, Copenhagen.

Foreningsmøder (1888). *Meddelelser fra den Botaniske forening i Kjøbenhavn*. H. Hagerups Boghandel, Copenhagen.

Foucault, P. (1990). *Le pêcheur d'orchidées. Aimé Bonpland (1877–1858)*. Seghers Collection Étonnants Voyageurs, Paris.

Foucault, M. (2010). *La fonction politique de l'intellectuel. Dits et écrits II*. Gallimard, Paris.

Fox Maule, A. (2011a). A.S. Ørsted [Online]. Available at: https://biografiskleksikon.lex.dk/A.S._%C3%98rsted_-_botaniker [Accessed 12 September 2021].

Fox Maule, A. (2011b). Eug. Warming [Online]. Available at: https://biografiskleksikon.lex.dk/Eug._Warming [Accessed 12 September 2021].

Friis, O. (1929). *Jylland i dansk litteratur indtil Blicher*. Gyldendalske boghandel, Nordisk forlag, Copenhagen.

Funder, H. (2013). En historie om foranderlighed. Evolutionsteorien i Danmark 1860–1880 [Online]. Available at: https://tidsskrift.dk/historisktidsskrift/article/view/56007 [Accessed 8 May 2021].

Ganong, W.F. (1901). *A Laboratory Course in Plant Physiology Especially as a Basis for Ecology*. Henry Holt and Company, New York.

Garcia, R. (ed.) (2015). *De Darwin à Lamarck : Kropotkine biologiste 1910–1919*. ENS Éditions, Lyon.

Garreau, J. (1967). Les chemins de fer en Norvège. *Norois*, 56, 585–600.

Gayon, J. (1992). Wallace et Darwin : enjeux d'un désaccord. In *Darwin et l'après Darwin. Une histoire de l'hypothèse de sélection naturelle*, Gayon, J. (ed.). Éditions Kimé, Paris.

Geffroy, A. (1851). *Histoire des États scandinaves (Suède, Danemark, Norvège)*. Librairie de Hachette et Cie, Paris.

GFSE (1909). *The Geographical Journal*. The Royal Geographical Society, 34(4), 448–450.

Gimingham, C.H. (2003). The Smith brothers: Scottish pioneers of modern ecology. *Botanical Journal of Scotland*, 55(2), 287–297.

Glaziou, A.F.M. (1905–1906). Plantæe Brasiliæ centralis à Glaziou lectæ. Liste des plantes du Brésil central recueillies en 1861–1895. *Société botanique de France*, 52(3), 1–112.

Glick, T.F. and Shaffer, E. (2014). *The Literary and Cultural Reception of Charles Darwin in Europe*. Bloomsbury Academic, London.

Gohau, G. (1990). *Une histoire de la géologie*. Le Seuil, Paris.

Goodland, R.J. (1975). The tropical origin of Ecology: Eugen Warming's jubilee. *Oikos*, 26, 240–245.

Graebner, P. (1895). Studien über die norddeutsche Heide, Versuch einer Formationsgliederung. *Englers Botanische Jahrbücher*, 20(4), 500–654.

Graebner, P. (1896). Klima und Heide in Norddeutschlaud. *Naturwissenschaftliche Wochenschrift*, XI, 196–207.

Graebner, P. (1898). Gliederung der Westpreussischen Vegetationsformationen. *Schriften der Natureforschenden Gesellschaft in Danzig, Commissions-Verlag von Vilhelm Engelmann in Leipzig*, Danzig, IX, 43–74.

Grandchamp, P. (2009). Des leçons de géologie du Collège de France au Discours sur les révolutions de la surface du Globe : quatre étapes successives du cheminement intellectuel de Cuvier. *Travaux du Comité français d'histoire de la géologie (COFRHIGÉO)*, 3(23), 17–65.

Grisebach, A.R.H. (1838). Über den Einfluss des Klimas auf die Begränzung der natürlichen Floren. *Linnaea*, 12, 159–200.

Grisebach, A.R.H. (1875). *La végétation du globe d'après sa disposition suivant les climats, esquisse d'une géographie comparée des plantes*. J.B. Baillière et fils, Paris.

Grove, R.H. (1996). *Green Imperialism: Colonial Expansion, Tropical Island Edens and the Origins of Environmentalism*. Cambridge University Press, Cambridge.

Grove, R.H. (2013). *Les îles du Paradis. L'invention de l'écologie aux colonies 1600–1854*. La Découverte, Paris.

Guédès, M. (1969). La théorie de la métamorphose en morphologie végétale : des origines à Goethe et Batsch. *Revue d'histoire des sciences et de leurs applications*, 22(4), 323–363.

Gunthert, A. (1999). La conquête de l'instantané. Archéologie de l'imaginaire photographique en France (1841–1895). PhD Thesis, École des Hautes Études en Sciences Sociales, Paris.

Haeckel, E. (1866). *Generelle Morphologie der Organismen*. Reimer, Berlin.

Haeckel, E. (1874). *Histoire de la création des êtres organisés d'après les lois naturelles*. Reinwald, Paris.

Hagen, J.B. (1983). The development of experimental methods in plant taxonomy (1920–1950). *Taxon, Journal of the International Association for Plant Taxonomy*, 32(3), 406–416.

Hagen, J.B. (1986). Ecologists and taxonomists: Divergent traditions in twentieth-century plant geography. *Journal of the History of Biology*, 19, 197–214.

Hampe, E. (1879). *Enumeratio muscorum hactenus in provinciis Brasiliensibus Rio de Janeiro et São Paulo detectorum*. Hoste A.A.F. & filium, Universitatis Bibliopolas, Havana.

Hansen, S. (1888). Lagoa-Santa Racen. En anthropologisk undersogelse af jordfundne menneskelevninger fra Brasilianske huler. Med et Tillaeg om det Jordfundne Menneske fra Pontimelo Rio de Arrecifes, La Plata. *En Samling af Afhandlinger e Museo lundii*, 1, 1–37.

Helle, A.E. (1992). *Histoire du Danemark*. Hatier, Paris.

Hjermitslev, H.H. (2011). Protestant responses to darwinism in Denmark, 1859–1914. *Journal of the History of Ideas*, 72(2), 279–303.

Höck, F. (1892). *Begleitpflanzen der Buche Botanischen Centralblatt*. Verlag von Gübruder Gotthelft, Cassel.

Holdridge, L.R. (1947). Determination of word plant formations from simple climatic data. *Science*, 105(2727), 367–368.

Holdridge, L.R. (1953). *Curso de ecología tropical*. Unpublished manuscript.

Holdridge, L.R. (1967). *Life Zone Ecology*. Tropical Science Center, San José.

Holten, B., Sterll, M., Fjeldså, J. (2004). *Den forsvundne maler P.W. Lund og P.A. Brandt i Brasilien*. Museum Tusculanums Forlag, Copenhagen.

Holttum, R.E. (1911–1912). Régions arctiques. *Annales de Géographie*, 21(119), 309–313.

Holttum, R.E. (1922). The vegetation of West Greenland. *Journal of Ecology*, 10(1), 87–108.

Hulin, N. (2002). L'enseignement des sciences naturelles au XIXe siècle dans ses liens à d'autres disciplines. *Revue d'histoire des sciences*, 55(1), 101–120.

Hult, R. (1881). Försök till analytisk behandling af växtformationerna. *Meddelanden af Societas pro Fauna et Flora Fennica*, VIII, 1–155.

von Humboldt, A. (1807). *Essai sur la géographie des plantes*. Schoell et Tübingen, Cotta, Paris.

Jacobsen, J.P. and Møller, V. (1893). *Darwin Hans liv og hans lær*. Gyldendalske Boghandel/Nordisk Forlag (F. Hegel & søn), Copenhagen.

Jacobsen, J.C. and Rothe, T. (1879). *Description des serres du jardin botanique de l'Université de Copenhague, avec l'explication du plan du jardin, tel qu'il a été arrêté et exécuté en 1871–1874*. Imprimerie de Tiele, Copenhagen.

Jensen, C. (1885). Mosser fra Novaia Zemlia, samlede paa Dijmphna Expeditionen 1882–1883 af Th. Holm. *Dijmphna-Togtets zoologiske-botaniske udbytte*, Copenhagen.

Jensen, J.L. (1985). Romantisme et christianisme au Danemark. *Romantisme*, 50, 111–124.

Johannsen, W. (1909). *Elemente der Exakten Erblichkeitslehre*. Verlag von Gustav Fischer, Iéna.

Johannsen, W. (1911). The genotype conception of heredity. *The American Naturalist*, 45(531), 129–159.

Jolivet, A. (1960). *Les romans de Henrik Pontoppidan : cinquante années de vie danoise*. Bibliothèque nordique, Paris.

Jolivet, A. (ed.) (1961). *Le visiteur royal de Henrik Pontoppidan*. Rombaldi, Paris.

Julien, C. (ed.) (2000). *L'histoire du développement scientifique et culturel de l'humanité*. Éditions UNESCO, Paris.

Jumelle, H. (1890). Le laboratoire de biologie végétale de Fontainebleau dirigé par Gaston Bonnier. *Revue générale de botanique*, 2, 289–299.

Karsdal, B. (1995). Skolen gennem tiderne. Søndergaard S.M., Skolens bygningshistorie [Online]. Available at: https://www.ribekatedralskole.dk/om-skolen/om-skolens-historie/jubilaeumsskriftet-1995 [Accessed 13 August 2021].

Kerner von Marilaün, A. (1863). *Das Pflanzenleben der Donaülander*. Wagner, Innsbruck.

Kjærgaard, P.C., Bek-Thomsen, J., Brøndum, L., Grumsen, S., Hjermitslev, H.H., Jensen, G.V., Larsen, M., Thomasen, L.S. (2006). Darwin in Denmark – An introduction. In *The Complete Work of Charles Darwin,* van Wyhe, J. (ed.) [Online]. Available at: http://darwin-online.org.uk [Accessed 17 October 2021].

Klein, A.L. (2000). *Eugen Warming e o cerrado brasileiro*. Editora UNESP, São Paulo.

Klinge, J. (1890). Über den Einfluss der mittleren Windrichtung auf das Verwachen der Gewässer nebst Betrachtungen anderer von der Windrichtung abhängiger Vegetations-Erschenugen im Ostbaltikum. *Botanische Jahrbücher Fur Systematik, Pflanzengeschichte Und Pflanzengeographie*, 11, 264–313.

Koch, L.J. (1945). Ribe Katedralskole 1886–1892. Nogle Indtryk [Online]. Available at: https://tidsskrift.dk/fraribeamt/article/view/75681 [Accessed 14 April 2021].

Kohler, H.C. (1989). Geomorfologia carstica na regiâo de Lagoa Santa – Minas Gerais. PhD Thesis, Institut de Géographie de São Paulo, São Paulo.

Kolderup Rosenvinge, L. (1923–1924). Eug. Warming. Tale i Videnskabernes Selskabs Mode 9. Maj 1924. *Bulletin de l'Académie royale des sciences et des lettres du Danemark*, 121–126.

Kolderup Rosenvinge, L. and Warming, E. (eds) (1912–1918, 1918–1920, 1930–1945). *The Botany of Iceland*. John Wheldon & Company, Copenhagen.

Köppen, W.P. (1900). Versuch einer Klassifikation der Klimate vorzugsweise nach ihren Beziehungen zur Pflanzen welt. *Geographical Zoology*, 6, 593–611.

Köppen, W.P. and Geiger, R. (1930). *Handbuch der Klimatologie*. Gebrüder Borntraeger, Berlin.

Korschinsky, S.I. (1901). Heterogenesis und Evolution. Ein Beitrag zur Théorie der Entstehung der Arten. *Flora oder Allgemeine Botanische Zeitung*, 240–368.

Krebs, G. (1984). Universités, professeurs et étudiants en Allemagne dans la première moitié du XIXe siècle. In *Aspects du Vormärz : Société et Politique en Allemagne dans la première moitié du XIXe siècle*, Krebs, G. (ed.). Presses Sorbonne Nouvelle, Paris.

Kropotkine, P. (1910). The direct action of environment on plants. *The Nineteenth Century and After*, 68(301), 58–77.

Kuhlmann, M. (1947). *Como Herborizar Material Arboreo*. Instituto de Botanica, São Paulo.

Kuhn, T.S. (1962). *The Structure of Scientific Revolutions*. Flammarion, Paris.

Kury, L. (2001). *Histoire naturelle et voyages scientifiques (1780–1830)*. L'Harmattan, Paris.

Lachmann, P. (1884–1885). Notice sur le jardin botanique de Buitenzorg dans l'île de Java. *Annales de la Société botanique de Lyon*, 12, 1–15.

Lahondère, C. (1997). Initiation à la phytosociologie sigmatiste. *Bulletin de la Société Botanique du Centre-Ouest*, 16, 47.

Lahondère, C. (2004). Les salicornes sur les côtes françaises. *Bulletin de la Société botanique du Centre-Ouest*, 24.

Laissus, Y. (ed.) (1995). *Les naturalistes français en Amérique du Sud XVIe-XIXe siècles*. Éditions du CTHS, Paris.

Lamarck, J.-B. (1963). *Zoological Philosophy an Exposition with Regard to the Natural History of Animals*. Hafner Publishing Company, New York/London.

Larguèche, A. (2015). *Christiana (1811–1869) : littérature, nation et religion dans la Norvège des années romantiques*. Presses Universitaires du Midi, Toulouse. doi: 10.4000/books.pumi.12312.

Lecoq, H. (1854–1858). *Étude sur la géographie botanique de l'Europe et en particulier sur la végétation du Plateau Central de la France*. Baillière, Paris.

Legros, J.-P. (2019). Les glaciations : découverte, étude, traces dans les paysages. *Bulletin de l'Académie des Sciences et Lettres*, 50(18).

Lehmann, E. (1931). Deux réformateurs du protestantisme danois : Kierkegaard et Grundtvig. *Revue d'histoire et de philosophie religieuses*, 11(6), 499–505.

Leine, K. (2018). *Rød mand/sort mand*. Gyldendal, Copenhagen.

Lenay, C. (1997). Yves Delage : évolution et hérédité d'un point de vue néolamarkien. In *Jean-Baptiste Lamarck 1744–1829*, Laurent, G. (ed.). Éditions du CTHS, Paris.

Lindeman, R. (1942). The trophic-dynamic aspect of ecology. *Ecology*, 23(4), 399–418.

von Linné, C. (1835). *Systema naturae per regna tria naturæ, secundum classes, ordines, genera, species*. Leyde.

Loison, L. (2010). *Qu'est-ce que le néolamarckisme ? Les biologistes français et la question de l'évolution des espèces*. Vuibert, Paris.

Loison, L. (2012). Le projet du néolamarckisme français (1880–1910). *Revue d'histoire des sciences*, 65(1), 61–79.

de Luna Filho, P.E. (2007). Peter Wilhelm Lund : O auge das suas investigações científicas e a razão para o término das pesquisas. PhD Thesis, Université de São Paulo, São Paulo.

Lund, P.W. (1835). Bemaerkninger over vegetation paa de indre hogsletter of Brasilien, isaer i plantehistorisk henseende. *Videnskabers-Selskabs Skrivter*, 6, 145–188.

Lund, P.W. (1845–1849). Lettre adressée à C.C. Rafn du 28 mars 1844. Notice sur des ossements humains fossiles, trouvés dans une caverne du Brésil. *Mémoires de la Société Royale des Antiquaires du Nord*, 49–77.

Lund, S. (1883). Notitser fra Universitetets botaniske Have. *Meddelelser fra den Botaniske forening*. Hagerups Boghandel, Copenhagen, 38–58.

Lyell, C. (1830–1833). *Principles of Geology, Being an Attempt to Explain the Former Changes of the Earth's Surface, by Reference to Causes now in Operation*. John Murray, London.

Mackenthun, G. (2016). Peter Wilhelm Lund, Universität Rostock. Institut für Anglistik/Amerikanistik [Online]. Available at: https://www.iaa.uni-rostock.de/forschung/laufende-forschungsprojekte/american-antiquities-prof-mackenthun/project/agents/peter-wilhelm-lund/ [Accessed 13 September 2022].

Magnin-Gonze, J. (2004). *Histoire de la botanique*. Delachaux et Niestlé, Paris.

Margalef, R. (1968). *Perspectives on Ecological Theory*. University of Chicago Press, Chicago.

Marsh, G.P. (1864). *Man and Nature, or Physical Geography as Modified by Human Action*. Sampson Low, London.

Martins, C. (1840). Observations sur les glaciers du Spitzberg, de la Suisse et de la Norvège. *Bibliothèque universelle de Genève*, 36.

Martins, C. (1843a). Remarques et expériences sur les glaciers sans névé de la chaine du Faulhorn. *Bulletin de la Société géologique de France*, 14, 133–145.

Martins, C. (1843b). Un hivernage scientifique en Laponie. *La Revue indépendante*, 31.

Martins, C. (1859). Vegetable colonization of British Isles, of Shetlands, Faroe and Iceland. Report, Annual report of the board of regents of the Smithsonian Institution, showing the operations, expenditures, and conditions of the institution for the year 1858, 229–237.

Martius, K.F.P., Eicher, A.W. von Müller, J., Projet, A. (1840–1906). *Flora Brasiliensis enumeratio plantarum in Brasilia hactenus detectarum : Quas suis aliorumque botanicorum studiis descriptas et methodo naturali digestas partim icone illustratas*. Monachii et Lipsiae, Munich/Leipzig.

Massart, J. (1912). *Pour la protection de la nature en Belgique*. Libraire Éditeur, Brussels.

Matagne, P. (2004). Les espèces sociales et leurs milieux ou l'écologie sociale balzacienne. In *Balzac géographe, territoires*, Dufour, P. and Mozet, N. (eds). Christian Pirot Éditeurs, Saint-Cyr-sur-Loire.

Matagne, P. (2009). *La naissance de l'écologie*. Ellipse, Paris.

Matagne, P. (2014). Les niveaux d'organisation : enjeux épistémologiques pour l'ingénierie écologique. In *Ingénierie écologique. Action par et/ou pour le vivant ?*, Rey, F., Gosselin, F., Doré, A. (eds). Éditions Quae, Versailles.

Matagne, P. (2016). Géographie-écologie occasions manquées et opportunités. In *Manifeste pour une géographie environnementale*, Chartier, D. and Rodary, E. (eds). SciencesPo, Paris.

Matagne, P. (ed.) (2022). *La nature en ville : Sociétés savantes et pratiques naturalistes (XIXe–XXesiècles)*. LISAA Éditeur, Champs sur Marne [Online]. Available at: http://books.openedition.org/lisaa/1668 [Accessed 28 November 2022].

Mathis, C.F. (2010). *In Nature We Trust. Les paysages anglais à l'ère industrielle*. Presses de l'Université Paris-Sorbonne, Paris.

McDougal, D.T. (1902). The origins of species by mutation. *Torreya, Torrey Botanical Society*, 2(7), 97–101.

McIntosh, R. (1985). *The Background of Ecology. Concepts and Theory*. Cambridge University Press, Cambridge.

Mendel, G. (1865). Versuche über Pflanzenhybriden. *Verhandlungen des naturforschenden Vereines*, 4, 3–47.

Merriam, C.H. (1892). The geographic distribution of life in North America with special reference to the Mammalia. *Proceedings of the Biological Society of Washington*, 7, 1–64.

Merriam, C.H. (1895). The geographic distribution of animals and plants in North America. US Department of Agriculture, Smithsonian Institution, Washington, 203–214.

Mikkelsen, E. (1913). *Perdus dans l'Arctique, récit de l'expédition de l'Alabama 1909–1912*. Imprimerie Mame et fils, Tours.

Millan, M. (2016). Analyse de la variabilité des traits architecturaux des formes de croissance dans les communautés végétales. PhD Thesis, Université de Montpellier, Montpellier.

Molinier, R. and Vignes, P. (1971). *Écologie et biocénotique*. Delachaux et Niestlé, Neuchâtel.

Möller, H.J. (1899). Clapodus nymani n gen, n sp., eine Podostemaceae aux Java. *Annales du Jardin botanique de Buitenzorg*, 16(2), 115–132.

Moquin-Tandon, A. (1841). *Éléments de tératologie végétale ou histoire des anomalies dans l'organisation des végétaux*. P.-J. Libraire Éditeur, Paris.

Moreira, N.C. and Stehmann, J.R. (2020). Eugen Warming's Florula Lagoensis revisited: Old lessons to new challenges. *Plant Ecology and Evolution*, 153(1), 143–151.

Morin, E. (2021). *Leçons d'un siècle de vie*. Denoël, Paris.

Mougel, F.-C. (2006). *L'Europe du Nord contemporaine de 1900 à nos jours*. Ellipse, Paris.

Muenscher, W.C. (1922). The effect of transpiration in the absorption of salts by the plants. PhD Thesis, Cornell University, Ithaca.

Musset, L. (1948). Les villes du Danemark. Origines et évolution. *Annales de Géographie*, 57(308), 308–321.

Nathorst, A.G. (1892). Kritische Bemerkungen über die Geschichte der Vegetation Grönlands. *Englers Botanische Jahrbücher*, 14, 183–221.

Neumann, R.P. (2017). Life zones: The rise and decline of a theory of the geographical distribution of species. In *Spacializing the History of Ecology*, de Bont, R. and Lachmund, J. (eds). Routledge, New York.

Neuvième Congrès international de géographie (1908). Le Globe. *Revue genevoise de géographie*, 47, 63–116.

Nicolson, M. (2013). Community concepts in plant ecology: From Humboldtian plant geography to the superorganism and beyond. *Web Ecology*, 13. doi: 10.5194/we-13-95-2013.

Nielsen, N. (1938). Nogle Bemaerkninger om Marskdannelsen i det dansk Vadehav. *Geografisk Tidsskrift*, 41, 123–138.

Nielsen, J.K. and Helama, S. (2012). Christian Theodor Vaupell, a Danish 19th century naturalist and a pioneering developer of the Quaternary geoscience. *History of Geo and Space Sciences*, 3, 143–150.

Nordbrandt, H. (1986). *Håndens skælven i november*. Digte, Brøndum.

Offner, J. (1930). Associations végétales du globe, d'après le professeur Eduard Rübel. *Annales de Géographie*, 39(222), 628–631.

Oorschot, L. (2009). The Flooded Village of Mando [Online]. Available at: http://www.mandoe.eu/The_Flooded_Village_of_Mando.pdf [Accessed 12 November 2021].

Ørsted, A.S. (1858). *Centralamericas gesneraceer et systematisk, plantegeographisk Bidrag til Centralamericas Flora*. Trykt Bianco Lunos Bogtrykkeri ves F.S. Muhle, Copenhagen.

Ørsted, A.S. (1863). *L'Amérique Centrale : recherches sur sa flore et sa géographie physique. Résultats d'un voyage dans les états de Costa Rica et de Nicaragua exécuté pendant les années 1846–1848*. Imprimerie de Bianco Lunos par F.S. Muhle, Copenhagen.

Ørsted, A.S. (1868). *Den tilbageskridende Metamorfose som normal Udviklingsgang, naermest Hensyn til Gymnospermernes Blomster*. Herman H.J. Lynge & Søn, Copenhagen.

Ostenfeld, C.H. and Hansen, C. (1925). *Flowering plants and ferns from North-Western Greenland collecting during the jubilee expedition 1920–1922 and some remarks on the vegetation of North Greenland. Arbejder fra den Botaniske have i København, Københavns universitet. Botanisk have*. Reitzel, Copenhagen.

de Oviedo and Valdés, G.F. (1851–1855). *Historia General y natural de las Indias*. Imprenta de la Real Academia de la Historia, Madrid.

Paludan, A.M. (1995). *Træk af elevernes historie. Ribe Byes forenede Klub-selskab ca., 1800–1840* [Online]. Available at: https://www.ribekatedralskole.dk/om-skolen/om-skolens-historie/jubilaeumsskriftet-1995 [Accessed 13 August 2021].

Parmentier, P. (1893). La botanique systématique et les théories de M. Vesque. *Mémoires de la Société d'émulation du Doubs*, 6(8), 156–171.

Patou-Mathis, M. (2012). De la paléontologie du XIXe siècle à l'archéozoologie du XXe siècle. *Les nouvelles de l'archéologie*, 129, 29–35.

Paul-Dubois, L. (1909). Grundtvig et le relèvement du Danemark au XIXe siècle. *Revue des Deux Mondes*, 52, 657–676.

Paulsen, O., Thorvald Sørensen T., Grøntved, J. (1942). *The Botany of Iceland*. Einar Munksgaard, Copenhagen.

Pavillard, J. (1901). *Éléments de biologie végétale*. Société d'éditions scientifiques, Montpellier, Paris.

Pavillard, J. (1912). Essai sur la nomenclature phytogéographique. *Bulletin de la Société languedocienne de géographie*, 5, 165–176.

Pavillard, J. (1918). Les progrès de la nomenclature dans la géographie botanique. *Annales de Géographie*, 27(150), 401–415.

Pavillard, J. (1935). *Éléments de sociologie végétale (phytosociologie)*. Hermann & Cie, Paris.

Peder Heiberg, P. (1866). *Botanisk Tidsskrift udgivet af den botaniske forening I Kjøbenhavn*. H. Hagerups Forlag, Copenhagen.

Pedersen, C.B. (2019). L'histoire hermaphrodite. Sur les Groenlands de Kim Leine. *Nordiques. Revue biannuelle*, 37, 57–75.

Penck, A. and Brückner, E. (1909). *Die Alpen im Eiszeitalter*. Tauchnitz, Leipzig.

Percy Groom, J.B.F. (1932). Obituary notices of fellows of the royal society 1865–1931. *The Royal Society*, 1(1), 63–64.

Pereira de Lucena, U. and Campolina de Sá Araújo, J. (2020). Lagoa Santa, contribuição para a geografia fitobiológica e a construção acadêmica do (os) cerrado (s): discussões de Peter Lund e Eugenius Warming sobre a origem do cerrado. *Brazilian Journal of development*, 6(12), 94165–94183.

Pérez Irungaray, G.E., Rosito Monzón, J.C., Maas Ibarra, R.E., Gándara Cabrera, G.A. (2018). *Ecosistemas de Guatemala basado en el sistema de clasificación de zonas de vida*. IARNA-URL, Instituto de Investigación y Proyección sobre Ambiente Natural y Sociedad de la Universidad Rafael Landívar, Guatemala.

Petersen, O.G. (1887–1891). *Meddelelser fra den Botaniske*. 45–47.

Petersen, E.S. (1972). "Foreningen af 5 Oktober 1898" og den nationale jordkamp i Nordslesvig. *Sønderjyske Årbøger Udgivet af Historisk Samfund for Sønderjylland*, 84, 80–146.

Poeppig, E.F. and Endlicher, I.L. (1835). *Nova genera ac species plantarum, quas in regno Chilensi Peruviano et in terra Amazonica*. Sumptibus Friderici Hofmeister, Leipzig.

Pokorny, A. (1871). De l'origine des plantes alpines. *Schriften des Vereins zur Verbreitung naturwissenschaftlicher, Kentnisse in Wien*, 7, 15.

Pontoppidan, E. (1763–1781). *Den Dansk Atlas*. Continued by Hans de Hofman, Jacob Langebek, Betthel Christian Sandvig.

Pontoppidan, H. (1926). *Ilum Galgebakke ; Den første Gendarm ; Nattevagt*, 1st edition. Vilhelm Øhlenschläger, Gyldendal/Dansklærerforeningen, Copenhagen.

Poreau, B. (2014). Biologie et complexité : histoire et modèles du commensalisme. PhD Thesis, Université Claude Bernard Lyon I, Lyon.

Poulsen, B. (2016). *Global Marine Science and Carlsberg – The Golden Connections of Johannes Schmidt (1877–1933)*. Brill, Leiden, Boston.

Pound, R. and Clements, F.E. (1898–1900). *The phytogeography of Nebraska*. Lincoln, Nebraska.

Prytz, S. (1984). *Warming Botaniker og Rejsende*. Bogan, Lynge.

Raby, M. (2017). *American Tropics: The Caribbean Roots of Biodiversity Science*. The University of North Carolina Press, Chapel Hill.

Raumolin, J. (1984). L'homme et la destruction des ressources naturelles : la Raubwirtschaft au tournant du siècle. *Annales Économies, Sociétés, Civilisations*, 39(4), 798–819.

Raunkiær, C. (1889). Vesterhavets Øst-og Sydkysts Vegetation. 1. *Festskrift i Anledning af Borchs Kollegiums 200-Aars Jubilæum*, 11, 317–362.

Raunkiær, C. (1904). Om biologiske typer, med Hensyn til Planternes Tilpasning til at overle ugunstige Aarister. *Botanisk Tidsskrift*, 26(14).

Raunkiær, C. (1905). Types biologiques pour la géographie botanique. *Académie Royale des Sciences et des Lettres du Danemark. Bulletin de l'année 1905*, 5, 347–437.

Raunkiær, C. (1934). *The Life Forms of Plants and Statistical Plant Geography, Being the Collected Papers of C. Raunkiær*. Clarendon Press, Oxford.

Rauser, F. (2004). L'éducation au Danemark. *Revue internationale d'éducation de Sèvres*, 36 [Online]. Available at: http://journals.openedition.org/ries/1550 [Accessed 8 May 2022].

Reclus, É. (1890). *Nouvelle Géographie Universelle. La terre et les hommes. XV, L'Amérique boréale*. Librairie Hachette et Cie, Paris.

Reidenbaugh, T.G. and Banta, W.C. (1980). Origin and effects of Spartina wrackina virginia salt Marsh. *Gulf Research Reports*, 6(4), 393–401.

Reiter, H. (1885). *Die Consolidation der Physiognomik, Als versuch einer Oekologie der Gewaechse*. Leuschner & Lubensky, Graz.

Renault, E. (2002). *Philosophie chimique : Hegel et la science dynamiste de son temps*. Presses Universitaires de Bordeaux, Pessac.

Rhein, C. (2003). L'écologie humaine, discipline-chimere. *Sociétés contemporaines*, 1–2(49–50), 167–190 [Online]. Available at: https://www.cairn.info/revue-societes-contemporaines-2003-1-page-167.htm [Accessed 22 September 2022].

Riise, F. (1924). *Danmarks Deltagelse i Verdensudstillingen i Rio de Janeiro 1922–1923*. Beretning afgivet til Kommissariatet, G.E.C. Gad i komm, Copenhagen.

Rivet, A. (1908). La Race Lagoa-Santa chez les populations précolombiennes de l'Équateur. *Bulletins et Mémoires de la Société d'Anthropologie de Paris*, 5(9), 209–274.

de Rothmaler, A. (1921). Les Hautes-Écoles de paysans au Danemark. *La revue pédagogique*, 79, 189–208.

Rouillé, A. and Marbot, B. (1986). *Le corps et son image : photographies du XIXe siècle*. Contrejour, Paris.

Roux, F. (2022). Les raisonnements des premiers phytosociologues (1910–1940) : convergences épistémologiques avec les sciences sociales. PhD Thesis, Centre François Viète, Brest.

Royal Danish Academy of Sciences and Letters (1896). *Oversigt over det Kongelige Danske Videnskabernes Selskabs Forhandlinger*. Bianco Lunos Kgl. of-Bogtrykkeri (F. Dreyer), Copenhagen.

Royal Danish Academy of Sciences and Letters (1898–1899). *Oversigt over det Kongelige Danske Videnskabernes Selskabs Forhandlinger.* Selskabets Medlemmer ved Begyndelsen af Aaret 1898, Bianco Lunos Kgl. of-Bogtrykkeri (F. Dreyer), Copenhagen.

Rübel, E. (1930). *Pflanzengesellschaften der Erde.* Verlag Hans Huber, Bern/Berlin.

Rubinstein, D. and Confino, M. (1992). Kropotkine savant. *Cahiers du monde russe et soviétique*, 33(2–3), 243–301.

Sachs, J. (1865). *Handbuch der Experimental-Physiologie der Pflanzen.* Verlag von Vilhelm Engelmann, Liepzig.

Sahli, R. (2016–2017). Étude phytochimique de quelques plantes extrêmophiles tunisiennes et exploration de leurs activités physiologiques. PhD Thesis, Université du Droit de la Santé Lille II/Université de Carthage, Lille/Carthage.

de Saint-Hilaire, A. (1837). *Tableau géographique de la végétation primitive dans la province de Minas Gerais.* A. Pihan de la Forest, Paris.

Sawyer, J.O. and Lindsey, A.A. (1963). The holdridge bioclimatic formations of the eastern and central United States. *Indiana Academy of Science*, 72, 105–112.

Scheuner, U. (2016). L'État, la formation et le contrôle des associations dans l'Allemagne du XIXe siècle. *Trivium.* doi: 10.4000/trivium.5288.

Schimper, K.F. (1837). Ueber die Eiszeit (Auszug aus einem Brief an L. Agassiz). *Actes de la Société Helvétique des Sciences Naturelles*, 22, 38–51.

Schimper, A.F.W. (1890). Über die Schutzmittel des Laubes gegen Transpiration. *Sitzungsberichte der Preussischen Akademie der Wissenschaften Math. – physik*, 2, 1045–1062.

Schimper, A.F.W. (1891). *Die indo – malayische Strandflora.* Verlag von Gustav Fischer, Iéna.

Schimper, A.F.W. (1898). *Pflanzen-Geographie, auf physiologischer grundlage.* Verlag von Gustav Fischer, Iéna.

Schimper, A.F.W. (1903). *Plant Geography upon a Physiological Basis.* Clarendon Press, Oxford.

Schmidt, J. (1899–1900). *Flora of Koh Chang: Contribution to the Knowledge of the végétation in the Gulf of Siam.* B. Luno, Copenhagen.

Schnell, R. (1960). Techniques d'herborisation et de conservation des plantes dans les pays tropicaux. *Journal d'agriculture tropicale et de botanique appliquée*, 7(1-3), 1–48.

Schou, R. (1900). *Om Landbruget Danmark – Agriculture en Danemark.* Librairie agricole de la Maison Rustique, Paris.

Septième Congrès international de géographie (1900). *Le Globe. Revue genevoise de géographie.*

Stebler, F.G. and Schröter, C. (1892). Beiträge zur Kenntnis der Matten und Weiden der Schweiz X: Versuch einer Übersicht über die Wiesentypen der Schweiz. *Landwirtschaftliches Jahrbuch der Schweiz*, 6, 95–212.

Simon, L. (1998). *Les paysages végétaux. Synthèse géographique*. Armand Colin, Paris.

Simon, L. (2000). Les forêts de plaine en Europe : ecologie, dynamisme et faits de répartition. *L'information géographique*, 64(1), 37–57.

Smith, W.G. (1914). Notes on Danish vegetation. *Journal of Ecology*, 2(2), 65–70.

Solms-Laubach, H. (1905). *Die leiteiden Gesichtspunkte einer allgemeinen Pflanzengeographie in kurzer Darstellung*. A. Félix, Leipzig.

Sørensen, T. and Grøntved, J. (1949). *The Botany of Iceland*. Einar Munksgaard/Oxford University Press, Copenhagen, Oxford.

Stangerup, H. (1983). *Lagoa Santa – Vidas e Ossadas, Rio de Janeiro*. Édition Nórdica LTDA, Rio de Janeiro.

Steenstrup, J. (1842). *Geognostik – geologisk Undersögelse af Skovmoserne Vidnesdam – og Lillemose i det nordlige Själland, ledsaget afsammenlignende Bemärkninger, hentede fra Danmarks Skov- Kjär- og Lyngmoser i Almindelighed*. Danske Videnskabernes Selskabs, Copenhagen.

Steinar, A.S. (2015). Making sense of a minor migrant stream. In *Expectations Unfulfilled: Norwegian Migrants in Latin America, 1820–1940*, Steinar, A.S. (ed.). Brill, Leiden.

Svabo, J.C. (1785). Indberetninger, indhentede paa en allernaadigst befalet, Reise i Færøe i Aarene 1781 og 1782. Rapport d'un voyage aux Féroé en 1781 et 1782.

Tansley, A.G. (1904). The problems of ecology. *The New Phytologist*, 3(8), 191–200.

Tansley, A.G. (1909). Review of the oecology of plants and introduction to the study of plant communities. *The New Phytologist*, 8(5/6), 218–227.

Tansley, A.G. (1911). *Types of British Vegetation*. Cambridge University Press, Cambridge.

Tansley, A.G. (1913). Review of a universal classification of plant-communities. *Journal of Ecology*, 1(1), 27–42.

Tansley, A.G. (1920). The classification of vegetation and the concept of development. *Journal of Ecology*, 8(2), 118–149.

Tansley, A.G. (1935). The use and abuse of vegetational concepts and terms. *Ecology*, 16(3), 284–307.

Tansley, A.G. (1947). The early history of modern plant ecology in Britain. *Journal of Ecology*, 35(1–2), 130–137.

The Royal Geographical Society (1885). The Danish Scientific Expedition of 1884 in the Gunboat "Fylla". In *Proceedings of the Royal Geographical Society and Monthly Record of Geography*, 7(3), 177–180.

Trimen, H. (1893). *Handbook of the Flora of Ceylon: Containing Descriptions of all the Species of Flowering Plants Indigenous to the Island, and Notes on their History, Distribution, and Uses: With an Atlas of Plates Illustrating some of the more Interesting Species*. Dulau and Co, London.

Trudgill, S. (2012a). Nature's clothing and spontaneous generation? The observations of Thoreau and Dureau de la Malle on plant succession. *Progress in Physical Geography*, 36(5), 707–714.

Trudgill, S. (2012b). The interactions between all neighbouring organisms: The roles of Charles Darwin, Ernst Haeckel and Eugenius Warming in the evolution of ideas on plant dynamics. *Progress in Physical Geography*, 36(6), 853–861.

Valdemar Sejrs Jordebog (1231). *Liber Census Daniæ*. Manuscript.

Vallin, J. (1989). La mortalité en Europe de 1720 à 1914 : tendances à long terme et changements de structure par sexe et par âge. Le déclin de la mortalité. *Annales de démographie historique*, 31–54.

Vaupell, C. (1851). *De nordsjællandske Skovmoser. En botanisk-mikroskopisk Undersøgelse af de Plantedele, som danne Tørven og af de Levninger af Fortidens Skove, der ere bevarede i nogle nordsjællandske Skovmoser*. C.A. Reitzel, Copenhagen.

Vaupell, C. (1858). De l'invasion du hêtre dans les forêts du Danemark. *Annales des Sciences Naturelles : Botanique*, 4(7), 55–86.

Verlot, B. (1865). *Le guide du botaniste herborisant*. J.B. Baillière et fils, Paris.

Vesque, J. (1878). Développement du sac embryonnaire des phanérogames Angiospermes. *Annales des Sciences Naturelles : Botanique*, 6(6), 238–285.

Vesque, J. (1882). L'espèce végétale considérée au point de vue de l'anatomie comparée. *Annales des Sciences naturelles : Botanique*, 6(13), 5–46.

Vesque, J. (1883). Contributions à l'histologie systématique de la feuille des caryophyllinées, précédées de remarques complémentaires sur l'importance des caractères anatomiques en botanique descriptive. *Annales des sciences naturelles : Botanique*, 6(15), 105–148.

Vesque, J. (1889). De L'emploi des caractères anatomiques dans la classification des végétaux. *Bulletin de la Société Botanique de France*, 36(10), XLI–LXXXIX.

Videnskabelige Expeditioner (1891). *Meddelelser fra den Botaniske forening i Kjøbenhavn*. H. Hagerups Boghandel, Copenhagen.

Viegnes, M., Jeanneret, S., Traglia, L. (eds) (2020). *Les lieux du polar. Entre cultures nationales et mondialisation*. Éditions Livreo-Alphil, Neuchâtel.

Vignaud, H. (1922). Le problème du peuplement initial de l'Amérique et de l'origine ethnique de la population indigène. *Journal de la Société des Américanistes*, 14–15, 1–63.

de Vries, H. (1905). *Species and Varieties, their Origin by Mutation*. Daniel Trembly MacDougal, Chicago/London.

de Vries, H. (1909). *Mutation Theory. Experiments and Observations on the Origins of Vegetable Kingdom*. The Open Court Publishing Company, Chicago.

Warburg, O. (1901). Vorschläge für die Einsführung einer geichmäßigen Nomenclatur in der Pflanzen-Geographie. *Botanische Jahrbücher fur Systematik Pflanzengeschichte und Pflanzengeographie*, 3(66), 23–30.

Warming, E. (1864). *Om lysfaenomener i planteriget. Tidsskrift for populære Fremstillinger af Naturvidenskaben*. P.G. Philipsens forlag, Copenhagen.

Warming, E. (1867, 1868). *Skildringer af naturen i det tropiske Brasilien navnlig i camposegnene. Tidsskrift for populære Fremstillinger af Naturvidenskaben*. P.G. Philipsens forlag, Copenhagen.

Warming, E. (1867–1893). *Symbolae ad floram Brasiliae centralis cognoscendam*. Naturhistoriske Forening i Kjøbenhavn, Copenhagen.

Warming, E. (1870a). *Kampen for Tilværelsen blandt Planterne. Tidsskrift for populære Fremstillinger af Naturvidenskaben*. P.G. Philipsens forlag, Copenhagen.

Warming, E. (1870b). Über die Entwicklung des Blütenstandes von Euphorbia. *Flora oder allgemeine botanische Zeitung*, 53, 385–397.

Warming, E. (1871). Er Koppen hos Vortemælken (Euphorbia L.) en Blomst eller en Blomsterstand ? En organogenetisk morfologisk Undersøgelse. Udgiven som afhandling for den filosofiske doktorgrad ved Kjøbenhavns universitet, Med tre kobbertavler, 8, Copenhagen, Leipzig. PhD Thesis, Bianco Lunos Bogtrykkeri, Copenhagen.

Warming, E. (1873). Untersuchungen über pollenbildende Phyllome und Kaulome. *Botanische Abhandlungen aus dem Gebiet der Morphologie und Physiologie*, 2(2), 1–90.

Warming, E. (1876). Die Blüte der Kompositen. *Botanische Abhandlungen*, 3(2), 1–167.

Warming, E. (1877). *Undersøgelser og betragtninger over Cycadeerne*. Bianco Lunos Bogtrykkeri, Copenhagen.

Warming, E. (1877–1878, 1882, 1883, 1889, 1890, 1891, 1893). *Symbolæ floram Brasiliæ centralis cognoscedam. Videnskabelige meddelelser fra den Naturhistoriske forening i Kjöbenhavn*. Bianco Lunos Bogtrykkeri, Copenhagen.

Warming, E. (1878a). De l'Ovule. *Annales des sciences naturelles : Botanique et biologie végétale*, 6, 177–266.

Warming, E. (1878b). *Haandbog i den systematiske Botanik (nærmest til brug for universitets-studerende og lærere)*. Copenhagen. English translation (1895). *A Handbook of Systematic Botany*. Swan Sonnenschein & Co, MacMillan & Co, London, New York.

Warming, E. (1880a). Über einige in den letzten Jahren gewonnenen Resultate in der Erforschung der Flora von Grönland. In *Botanische Jahrbücher für Systematik Pflanzengeschichte und Pflanzengeographic*, Engler, A. (ed.). Verlag von Wilhelm Engelmann, Leipzig.

Warming, E. (1880b). *Den almindelige Botanik: En Lærebog, nærmest til Brug for Studerende og Lærere*. P.G. Philipsens Forlag, Copenhagen.

Warming, E. (1880–1881). *Den Dansk botanisk literatur fra de ældste tider til 1880. Botanisk Tiddsskrift udgivet af den botaniske forening i Kjøbenhavn*. Carl Lunds Bogtrikkeri, Copenhagen.

Warming, E. (1882–1886). *Literatur-Anmeldelser. Meddelelser fra den Botaniske forening i Kjøbenhavn*. H. Hagerups Boghandel, Copenhagen.

Warming, E. (1884). *Om Skudbygning, Overvintring og Foryngelse*. Naturhistorisk Forenings Festskrift, Copenhagen.

Warming, E. (1888). *Om Grønlands vegetation, 1886–1887. Meddelelser om Grønland, udgivne af Commissionen for Ledelsen af de geologiske og geograpfiiske Undersøgelser i Grønland*. Tolvte Hefte, Copenhagen.

Warming, E. (1891). *Botaniske Exkursioner 1*. Fra Vesterhavskystens Marskegne. *Videnskabelige Meddelelser fra den Naturhistoriske Forening i Kjøbenhavn for Aaret 1890*. Bianco Lunos Kgl. of-Bogtrykkeri (F. Dreyer), Copenhagen.

Warming, E. (1892a). *Botaniske Exkursioner 2. De psammophile Formationer i Danmark. Videnskabelige Meddelelser fra den Naturhistoriske Forening i Kjøbenhavn for Aaret 1891*. Bianco Lunos Kgl. of-Bogtrykkeri (F. Dreyer), Copenhagen.

Warming, E. (1892b). *Lagoa Santa. Et Bidrag til den biologiske Plantegeografi. Det Kongelige Danske videnskabernes selskab Skrifter sjette Række, Naturvidenskabelig og mathematisk Afdeling*. Bianco Lunos Kgl. Hof-Bogtrykkeri, Copenhagen.

Warming, E. (1893a). Lagoa Santa (Brésil). Étude de Géographie Botanique. *Revue générale de botanique*, 145–158, 209–223.

Warming, E. (1893b). Ekskursionen til Fanø og Blaavand i Juli 1893. *Botanisk Tidsskrift*, 52–86.

Warming, E. (1895). *Plantesamfund Grundträk af den Ökologiske Plantegeografi*. P.G. Philipsen, Copenhagen.

Warming, E. (1896). *Lehrbuch der ökologischen Pflanzengeographie, eine Einführung in die Kenntniss der Pflanzenvereine*. Emil Knoblauch, Gebrüder Borntraeger, Berlin.

Warming, E. (1897). *Exkursioner 3. Skarridsø. Videnskabelige Meddelelser fra den Naturhistoriske Forening i Kjøbenhavn for Aaret 1897*. Bianco Lunos Kgl. of-Bogtrykkeri (F. Dreyer), Copenhagen.

Warming, E. (1899). On the vegetation of tropical America. *Botanical Gazette*, 27(1), 1–18.

Warming, E. (1900). *Plantelivet: Lærebog i Botanik pour Skoler og Seminarier*. Nordisk forlag, Copenhagen.

Warming, E. (1901) Om Bornholms Plantevækst. Den botaniske Studenter-Exkursion i 1901. *Botanisk Tidsskrift*, 281–353.

Warming, E. (ed.) (1901–1903–1908). *Botany of the Færöes based upon Danish investigations*. Det Nordiske Forlag/Wheldon & Co, Copenhagen, London.

Warming, E. (1904a). *Den Dansk Planteverdens historie efter istiden: Et Kortfattet overblik*. Trykt i Universitetbogtrykkeriet, Copenhagen.

Warming, E. (1904b). *Bidrag til Vadernes, Sandenes og Marskens Naturhistorie, Under Medarbejde Af Dr. Wesenberg-Lund, E. Østrup, etc*. Bianco Lunos Bogtrykkeri, Copenhagen.

Warming, E. (1906–1919). *Dansk Plantevækst*. 1. Gyldendalske Boghandel Nordisk Forlag, Copenhagen.

Warming, E. (1908). *Om Planterigets livsformer. Festskrift udgivet af*. Københavns Universitet, Copenhagen.

Warming, E. (ed.) (1908–1921). The structure and biology of Arctic flowering plants. *Meddelelser om Grønland*, 36–37.

Warming, E. (1909). *Oecology of Plants: An Introduction to the Study of Plant Communities*. Clarendon Press, Oxford.

Warming, E. (1913). *Descriptive Notes on the Topography and Vegetation of Some Localities Visited by the Excursion in Denmark Arranged for the Members of l'Association internationale des botanistes*. Dansk Botanik Forening Forlag, Copenhagen.

Warming, E. (1914). *Observations sur la valeur systématique de l'ovule. Mindeskrift i andledningaf hundredaaret for Japetus Steenstrups fødsel udgivet af en freds af Naturforskeve*. Ved Hector F.E. Jungersen og Eug. Warming Udgivelsen bekostet Carlsbergfondet, Copenhagen.

Warming, E. (1915). *Nedstamningslæren*. Udvalget til Folkeoplysnings Fremme, G.E.C. Gad, Copenhagen.

Warming, E. (1920). Caryophyllaceæ. *Meddelelser om Grønland*, 37, 229–342.

Warming, E. (1923). Økologiens Grundformer – Udkast til en systematisk Ordning. *Kongelige Danske Videnskabernes Selskabs Skrifter – Naturvidenskabelig og Mathematisk Afdeling*, 8(4), 120–187.

Warming, E. and Graebner, P. (1902). *Lehrbuch der ökologischen Pflanzengeographie, eine Einführung in die Kenntniss der Pflanzenvereine*. Verlag von Gebrüder Borntraeger, Berlin.

Warming, E. and Graebner, P. (1918). *Lehrbuch der ökologischen Pflanzengeographie*. Gebrüder Borntraeger, Berlin.

Weber, C.A. (1894). *Ueber Veränderungen in der Vegetation der Hochmoore unter dem Einflüss der Kultur mit Beziehung auf praktische Fragen. Mitteilungen des Vereins zur Förderung der Moorkultur im Deutschen Reiche*.

Weismann, A. (1893). *Das Keimplasma. Eine Theorie der Vererbung*. Walter Scott Ltd, London.

White, G. (1789). *The Natural History and Antiquities of Selborne in the County of Southampton. Histoire naturelle de Selborne*. Attitudes, Le mot et le reste, Paris.

Willdenow, K.L. (1805). *The Principles of Botany, and of Vegetable Physiology*. Blackwood, Edinburgh.

Winge, O. and Winge, H. (1888). *En samling af afhandlinger. De i det indre Brasiliens Karlkstenshuler af professor Dr Vilhelm Peter Lund, Lundske palaeontologiske Afdeling af Kjobenhavns, Indeholdende Afhandlinger af J. Reinhardt, Udgiveren og S. Hansen*. H. Hagerups Boghandel, Copenhagen.

Wojciech, R. (1987). Pharmaciens pionniers de la production de porcelaine. Le service "Flora Danica". *Revue d'histoire de la pharmacie*, 75(275), 319–324.

Worster, D. (1992). *Les pionniers de l'écologie. Une histoire des idées écologiques*. Éditions Sang de la Terre, Paris.

Wyss-Neel, H. (1971). *Le Jutland dans l'œuvre de Steen Steensen Blicher*. Presses universitaires de Caen, Caen [Online]. Available at: https://books.openedition.org/puc/1268 [Accessed 7 June 2022].

Yapp, R.H. (1922). The dovey salt marshes in 1921. *Journal of Ecology*, 10(1), 18–23.

Yapp, R.H., Johns, D., Jones, O.T. (1917). The salt marshes of the dovey estuary. *Journal of Ecology*, 5(2), 65–103.

Zeiger, E., Farquhar, G.D., Cowan, I.R. (eds) (1987). *Stomatal Function*. Standford University Press, Standford.

Index of Names

A, B

C, D

E, F

G, H

J, K

L, M

N, O

P, R

S, T

Index of Terms

E, F

G, H

L, M

N, O

P, S

T, Z

Other titles from

in

Ecological Science

2023

GAID Kader
Drinking Water Treatment 1: Water Quality and Clarification
Drinking Water Treatment 2: Chemical and Physical Elimination of Organic Substances and Particles
Drinking Water Treatment 3: Organic and Mineral Micropolluants
Drinking Water Treatment 4: Membranes Applied to Drinking Water and Desalination
Drinking Water Treatment 5: Calco-carbonic Equilibrium and Disinfection

2022

AMIARD Jean-Claude
Management of Radioactive Waste (Radioactive Risk Set – Volume 5)
Marine Radioecology (Radioactive Risk Set – Volume 6)

GODET Laurent, DUFOUR Simon, ROLLET Anne-Julia
The Baseline Concept in Biodiversity Conservation: Being Nostalgic or Not in the Anthropocene Era

GRISON Claude, CASES Lucie, LE MOIGNE Mailys, HOSSAERT-MCKEY Martine
Photovoltaism, Agriculture and Ecology From Agrivoltaism to Ecovoltaism

LÉVÊQUE Christian
Biodiversity Erosion: Issues and Questions

2021

FLEURENCE Joël
Microalgae: From Future Food to Cellular Factory

AMIARD Jean-Claude
Disarmament and Decommissioning in the Nuclear Domain
(Radioactive Risk Set – Volume 4)

JOLY Fernand, BOURRIÉ Guilhem
Mankind and Deserts 1: Deserts, Aridity, Exploration and Conquests
Mankind and Deserts 2: Water and Salts
Mankind and Deserts 3: Wind in Deserts and Civilizations

MUTTIN Frédéric, THOMAS Hélène
Marine Environmental Quality

PIOCH Sylvain, SOUCHE Jean-Claude
Eco-design of Marine Infrastructures: Towards Ecologically-informed Coastal and Ocean Development

2020

BRUSLÉ Jacques, QUIGNARD Jean-Pierre
Fish Behavior 1: Eco-ethology
Fish Behavior 2: Ethophysiology

LE FLOCH Stéphane
Remote Detection and Maritime Pollution: Chemical Spill Studies

MIGON Christophe, NIVAL Paul, SCIANDRA Antoine
The Mediterranean Sea in the Era of Global Change 1: 30 Years of Multidisciplinary Study of the Ligurian Sea
The Mediterranean Sea in the Era of Global Change 2: 30 Years of Multidisciplinary Study of the Ligurian Sea

ROSSIGNOL Jean-Yves
Climatic Impact of Activities: Methodological Guide for Analysis and Action

2019

AMIARD Jean-Claude
Industrial and Medical Nuclear Accidents: Environmental, Ecological, Health and Socio-economic Consequences
(Radioactive Risk Set – Volume 2)
Nuclear Accidents: Prevention and Management of an Accidental Crisis
(Radioactive Risk Set – Volume 3)

BOULEAU Gabrielle
Politicization of Ecological Issues: From Environmental Forms to Environmental Motives

DAVID Valérie
Statistics in Environmental Sciences

GIRAULT Yves
UNESCO Global Geoparks: Tension Between Territorial Development and Heritage Enhancement

KARA Mohamed Hichem, QUIGNARD Jean-Pierre
Fishes in Lagoons and Estuaries in the Mediterranean 2: Sedentary Fish
Fishes in Lagoons and Estuaries in the Mediterranean 3A: Migratory Fish
Fishes in Lagoons and Estuaries in the Mediterranean 3B: Migratory Fish

OUVRARD Benjamin, STENGER Anne
Incentives and Environmental Policies: From Theory to Empirical Novelties

2018

AMIARD Jean-Claude
Military Nuclear Accidents: Environmental, Ecological, Health and Socio-economic Consequences
(Radioactive Risk Set – Volume 1)

FLIPO Fabrice
The Coming Authoritarian Ecology

GUILLOUX Bleuenn
Marine Genetic Resources, R&D and the Law 1: Complex Objects of Use

KARA Mohamed Hichem, QUIGNARD Jean-Pierre
Fishes in Lagoons and Estuaries in the Mediterranean 1: Diversity, Bioecology and Exploitation

2016

BAGNÈRES Anne-Geneviève, HOSSAERT-MCKEY Martine
Chemical Ecology

2014

DE LARMINAT Philippe
Climate Change: Identification and Projections

Printed and bound by CPI Group (UK) Ltd, Croydon, CR0 4YY
18/03/2024
14472321-0005